Just The Facts 101
Textbook Key Facts

Dominican Republic Ecology & Nature Protection Handbook

by Cram101
Textbook NOT Included

Table of Contents

Title Page

Copyright

Foundations of Business

Management

Business law

Finance

Human resource management

Information systems

Marketing

Manufacturing

Commerce

Business ethics

Accounting

Index: Answers

Just The Facts101

Exam Prep for

Dominican Republic Ecology & Nature Protection Handbook

Just The Facts101 Exam Prep is your link from
the textbook and lecture to your exams.

**Just The Facts101 Exam Preps are unauthorized and comprehensive reviews
of your textbooks.**

All material provided by CTI Publications (c) 2019

Textbook publishers and textbook authors do not participate in or contribute to these reviews.

Just The Facts101 Exam Prep

Copyright © 2019 by CTI Publications. All rights reserved.

eAIN 444549

Foundations of Business

A business, also known as an enterprise, agency or a firm, is an entity involved in the provision of goods and/or services to consumers. Businesses are prevalent in capitalist economies, where most of them are privately owned and provide goods and services to customers in exchange for other goods, services, or money.

:: Meetings ::

An _____ is a group of people who participate in a show or encounter a work of art, literature, theatre, music, video games, or academics in any medium. _____ members participate in different ways in different kinds of art; some events invite overt _____ participation and others allowing only modest clapping and criticism and reception.

Exam Probability: **Medium**

1. *Answer choices:*

(see index for correct answer)

- a. Audience
- b. Stammtisch
- c. Conference hall
- d. Meeting point

Guidance: level 1

:: Data collection ::

A _____ is an utterance which typically functions as a request for information. _____s can thus be understood as a kind of illocutionary act in the field of pragmatics or as special kinds of propositions in frameworks of formal semantics such as alternative semantics or inquisitive semantics. The information requested is expected to be provided in the form of an answer. _____s are often conflated with interrogatives, which are the grammatical forms typically used to achieve them. Rhetorical _____s, for example, are interrogative in form but may not be considered true _____s as they are not expected to be answered. Conversely, non-interrogative grammatical structures may be considered _____s as in the case of the imperative sentence "tell me your name".

Exam Probability: **Medium**

2. *Answer choices:*

(see index for correct answer)

- a. PISCES
- b. Interpellation
- c. Dye tracing
- d. Question

Guidance: level 1

:: Employment ::

_____ is a relationship between two parties, usually based on a contract where work is paid for, where one party, which may be a corporation, for profit, not-for-profit organization, co-operative or other entity is the employer and the other is the employee. Employees work in return for payment, which may be in the form of an hourly wage, by piecework or an annual salary, depending on the type of work an employee does or which sector she or he is working in. Employees in some fields or sectors may receive gratuities, bonus payment or stock options. In some types of _____ , employees may receive benefits in addition to payment. Benefits can include health insurance, housing, disability insurance or use of a gym. _____ is typically governed by _____ laws, regulations or legal contracts.

Exam Probability: **Low**

3. *Answer choices:*

(see index for correct answer)

- a. Split shift
- b. Vanpool
- c. Work sharing
- d. Legal working age

Guidance: level 1

:: Survey methodology ::

An _____ is a conversation where questions are asked and answers are given. In common parlance, the word "_____" refers to a one-on-one conversation between an _____ er and an _____ ee. The _____ er asks questions to which the _____ ee responds, usually so information may be transferred from _____ ee to _____ er . Sometimes, information can be transferred in both directions. It is a communication, unlike a speech, which produces a one-way flow of information.

Exam Probability: **Low**

4. *Answer choices:*

(see index for correct answer)

- a. Coverage error
- b. National Health Interview Survey
- c. Census
- d. Data editing

Guidance: level 1

:: Contract law ::

A _____ is a legally-binding agreement which recognises and governs the rights and duties of the parties to the agreement. A _____ is legally enforceable because it meets the requirements and approval of the law. An agreement typically involves the exchange of goods, services, money, or promises of any of those. In the event of breach of _____ , the law awards the injured party access to legal remedies such as damages and cancellation.

Exam Probability: **Low**

5. *Answer choices:*

(see index for correct answer)

- a. Severable contract
- b. Perfect tender
- c. Contract
- d. Extended warranty

Guidance: level 1

:: Stock market ::

A shareholder is an individual or institution that legally owns one or more shares of stock in a public or private corporation. _____ may be referred to as members of a corporation. Legally, a person is not a shareholder in a corporation until their name and other details are entered in the corporation's register of _____ or members.

Exam Probability: **Medium**

6. *Answer choices:*

(see index for correct answer)

- a. Reverse stock split
- b. Shareholders

- c. Rogue trader
- d. Blue chip

Guidance: level 1

:: Accounting terminology ::

_____ is a legally enforceable claim for payment held by a business for goods supplied and/or services rendered that customers/clients have ordered but not paid for. These are generally in the form of invoices raised by a business and delivered to the customer for payment within an agreed time frame.
_____ is shown in a balance sheet as an asset. It is one of a series of accounting transactions dealing with the billing of a customer for goods and services that the customer has ordered. These may be distinguished from notes receivable, which are debts created through formal legal instruments called promissory notes.

Exam Probability: **Medium**

7. *Answer choices:*

(see index for correct answer)

- a. Capital appreciation
- b. Double-entry accounting
- c. Impairment cost
- d. Accounts receivable

Guidance: level 1

:: Human resource management ::

_____ is the corporate management term for the act of reorganizing the legal, ownership, operational, or other structures of a company for the purpose of making it more profitable, or better organized for its present needs. Other reasons for _____ include a change of ownership or ownership structure, demerger, or a response to a crisis or major change in the business such as bankruptcy, repositioning, or buyout. _____ may also be described as corporate _____ , debt _____ and financial _____ .

Exam Probability: **Low**

8. *Answer choices:*

(see index for correct answer)

- a. Attendance management
- b. Restructuring
- c. Pay in lieu of notice
- d. SLT Human Capital Solutions

Guidance: level 1

:: Health promotion ::

_____, as defined by the World _____ Organization, is "a state of complete physical, mental and social well-being and not merely the absence of disease or infirmity." This definition has been subject to controversy, as it may have limited value for implementation. _____ may be defined as the ability to adapt and manage physical, mental and social challenges throughout life.

Exam Probability: **High**

9. *Answer choices:*

(see index for correct answer)

- a. Health
- b. Patient navigators
- c. Health equity
- d. Health risk assessment

Guidance: level 1

:: National accounts ::

_____ is a monetary measure of the market value of all the final goods and services produced in a period of time, often annually. GDP per capita does not, however, reflect differences in the cost of living and the inflation rates of the countries; therefore using a basis of GDP per capita at purchasing power parity is arguably more useful when comparing differences in living standards between nations.

Exam Probability: **Low**

10. *Answer choices:*

(see index for correct answer)

- a. National Income
- b. Fixed capital
- c. capital formation

Guidance: level 1

:: Business process ::

A _____ or business method is a collection of related, structured activities or tasks by people or equipment which in a specific sequence produce a service or product for a particular customer or customers. _____ es occur at all organizational levels and may or may not be visible to the customers. A _____ may often be visualized as a flowchart of a sequence of activities with interleaving decision points or as a process matrix of a sequence of activities with relevance rules based on data in the process. The benefits of using _____ es include improved customer satisfaction and improved agility for reacting to rapid market change. Process-oriented organizations break down the barriers of structural departments and try to avoid functional silos.

Exam Probability: **Low**

11. *Answer choices:*

(see index for correct answer)

- a. Order processing
- b. Business process
- c. Desktop outsourcing
- d. Software Ideas Modeler

Guidance: level 1

:: Strategic alliances ::

A _____ is an agreement between two or more parties to pursue a set of agreed upon objectives needed while remaining independent organizations. A _____ will usually fall short of a legal partnership entity, agency, or corporate affiliate relationship. Typically, two companies form a _____ when each possesses one or more business assets or have expertise that will help the other by enhancing their businesses. _____ s can develop in outsourcing relationships where the parties desire to achieve long-term win-win benefits and innovation based on mutually desired outcomes.

Exam Probability: **High**

12. *Answer choices:*

(see index for correct answer)

- a. Cross-licensing
- b. International joint venture
- c. Defensive termination
- d. Strategic alliance

Guidance: level 1

:: ::

Culture is the social behavior and norms found in human societies. Culture is considered a central concept in anthropology, encompassing the range of phenomena that are transmitted through social learning in human societies. _____ universals are found in all human societies; these include expressive forms like art, music, dance, ritual, religion, and technologies like tool usage, cooking, shelter, and clothing. The concept of material culture covers the physical expressions of culture, such as technology, architecture and art, whereas the immaterial aspects of culture such as principles of social organization , mythology, philosophy, literature , and science comprise the intangible _____ heritage of a society.

Exam Probability: **Low**

13. *Answer choices:*
(see index for correct answer)

- a. functional perspective
- b. open system
- c. Character
- d. interpersonal communication

Guidance: level 1

:: ::

A _____ is an organization, usually a group of people or a company, authorized to act as a single entity and recognized as such in law. Early incorporated entities were established by charter. Most jurisdictions now allow the creation of new _____ s through registration.

Exam Probability: **Low**

14. *Answer choices:*

(see index for correct answer)

- a. open system
- b. personal values
- c. Corporation
- d. functional perspective

Guidance: level 1

:: International trade ::

_____ involves the transfer of goods or services from one person or entity to another, often in exchange for money. A system or network that allows _____ is called a market.

Exam Probability: **Medium**

15. *Answer choices:*

(see index for correct answer)

- a. Standard trading conditions
- b. Concertina model
- c. FAST Card
- d. Foreign trade multiplier

Guidance: level 1

:: Business law ::

A _____ is a group of people who jointly supervise the activities of an organization, which can be either a for-profit business, nonprofit organization, or a government agency. Such a board's powers, duties, and responsibilities are determined by government regulations and the organization's own constitution and bylaws. These authorities may specify the number of members of the board, how they are to be chosen, and how often they are to meet.

Exam Probability: **Medium**

16. *Answer choices:*

(see index for correct answer)

- a. Finance lease
- b. Board of directors
- c. Business valuation
- d. Ladenschlussgesetz

Guidance: level 1

:: Business planning ::

_____ is an organization's process of defining its strategy, or direction, and making decisions on allocating its resources to pursue this strategy. It may also extend to control mechanisms for guiding the implementation of the strategy. _____ became prominent in corporations during the 1960s and remains an important aspect of strategic management. It is executed by strategic planners or strategists, who involve many parties and research sources in their analysis of the organization and its relationship to the environment in which it competes.

Exam Probability: **High**

17. *Answer choices:*

(see index for correct answer)

- a. Joint decision trap
- b. operational planning
- c. Business war games
- d. Strategic planning

Guidance: level 1

:: Critical thinking ::

An _____ is a set of statements usually constructed to describe a set of facts which clarifies the causes, context, and consequences of those facts. This description of the facts et cetera may establish rules or laws, and may clarify the existing rules or laws in relation to any objects, or phenomena examined. The components of an _____ can be implicit, and interwoven with one another.

Exam Probability: **High**

18. *Answer choices:*

(see index for correct answer)

- a. Critical-Creative Thinking and Behavioral Research Laboratory
- b. Higher-order thinking
- c. Merseyside Skeptics Society
- d. Explanation

Guidance: level 1

:: Packaging ::

In work place, _____ or job _____ means good ranking with the hypothesized conception of requirements of a role. There are two types of job _____ s: contextual and task. Task _____ is related to cognitive ability while contextual _____ is dependent upon personality. Task _____ are behavioral roles that are recognized in job descriptions and by remuneration systems, they are directly related to organizational _____, whereas, contextual _____ are value based and additional behavioral roles that are not recognized in job descriptions and covered by compensation; they are extra roles that are indirectly related to organizational _____ . Citizenship _____ like contextual _____ means a set of individual activity/contribution that supports the organizational culture.

Exam Probability: **High**

19. *Answer choices:*

(see index for correct answer)

- a. Load securing
- b. Digipak
- c. Plain cigarette packaging
- d. Box-sealing tape

Guidance: level 1

:: Management ::

The term _____ refers to measures designed to increase the degree of autonomy and self-determination in people and in communities in order to enable them to represent their interests in a responsible and self-determined way, acting on their own authority. It is the process of becoming stronger and more confident, especially in controlling one's life and claiming one's rights. _____ as action refers both to the process of self- _____ and to professional support of people, which enables them to overcome their sense of powerlessness and lack of influence, and to recognize and use their resources. To do work with power.

Exam Probability: **Medium**

20. *Answer choices:*

(see index for correct answer)

- a. Lead scoring
- b. Managerial economics
- c. Innovation leadership
- d. Facilitator

Guidance: level 1

:: Systems theory ::

A _____ is a group of interacting or interrelated entities that form a unified whole. A _____ is delineated by its spatial and temporal boundaries, surrounded and influenced by its environment, described by its structure and purpose and expressed in its functioning.

Exam Probability: **High**

21. *Answer choices:*

(see index for correct answer)

- a. System
- b. equifinality
- c. Black box
- d. management system

Guidance: level 1

:: ::

_____ is the means to see, hear, or become aware of something or someone through our fundamental senses. The term _____ derives from the Latin word perceptio, and is the organization, identification, and interpretation of sensory information in order to represent and understand the presented information, or the environment.

Exam Probability: **Medium**

22. *Answer choices:*

(see index for correct answer)

- a. similarity-attraction theory
- b. Perception

- c. Character
- d. personal values

Guidance: level 1

:: International trade ::

> In finance, an _____ is the rate at which one currency will be exchanged for another. It is also regarded as the value of one country's currency in relation to another currency. For example, an interbank _____ of 114 Japanese yen to the United States dollar means that ¥114 will be exchanged for each US$1 or that US$1 will be exchanged for each ¥114. In this case it is said that the price of a dollar in relation to yen is ¥114, or equivalently that the price of a yen in relation to dollars is $1/114.

Exam Probability: **High**

23. *Answer choices:*
(see index for correct answer)

- a. Import quota
- b. Agreement on Government Procurement
- c. Spice trade
- d. Exchange rate

Guidance: level 1

:: Market research ::

A _____ is a small, but demographically diverse group of people and whose reactions are studied especially in market research or political analysis in guided or open discussions about a new product or something else to determine the reactions that can be expected from a larger population. It is a form of qualitative research consisting of interviews in which a group of people are asked about their perceptions, opinions, beliefs, and attitudes towards a product, service, concept, advertisement, idea, or packaging. Questions are asked in an interactive group setting where participants are free to talk with other group members. During this process, the researcher either takes notes or records the vital points he or she is getting from the group. Researchers should select members of the _____ carefully for effective and authoritative responses.

Exam Probability: **High**

24. *Answer choices:*

(see index for correct answer)

- a. Consumer neuroscience
- b. CoolBrands
- c. Brand elections
- d. 6-3-5 Brainwriting

Guidance: level 1

:: Financial risk ::

_____ is a type of risk faced by investors, corporations, and governments that political decisions, events, or conditions will significantly affect the profitability of a business actor or the expected value of a given economic action. _____ can be understood and managed with reasoned foresight and investment.

Exam Probability: **High**

25. *Answer choices:*
(see index for correct answer)

- a. Solvency cone
- b. Financial risk management
- c. Political risk
- d. Credit scorecards

Guidance: level 1

:: Stock market ::

_____ is freedom from, or resilience against, potential harm caused by others. Beneficiaries of _____ may be of persons and social groups, objects and institutions, ecosystems or any other entity or phenomenon vulnerable to unwanted change by its environment.

Exam Probability: **Low**

26. Answer choices:

(see index for correct answer)

- a. Cross listing
- b. Security
- c. Sector rotation
- d. London Stock Exchange Group

Guidance: level 1

:: Management ::

A _____ is a method or technique that has been generally accepted as superior to any alternatives because it produces results that are superior to those achieved by other means or because it has become a standard way of doing things, e.g., a standard way of complying with legal or ethical requirements.

Exam Probability: **Low**

27. Answer choices:

(see index for correct answer)

- a. Innovation leadership
- b. Supplier relationship management
- c. Industrial market segmentation
- d. Authoritarian leadership style

Guidance: level 1

:: Classification systems ::

_____ is the practice of comparing business processes and performance metrics to industry bests and best practices from other companies. Dimensions typically measured are quality, time and cost.

Exam Probability: **High**

28. *Answer choices:*

(see index for correct answer)

- a. Stellar classification
- b. Benchmarking
- c. Bliss bibliographic classification
- d. Systematized Nomenclature of Medicine

Guidance: level 1

:: Management occupations ::

_____ is the process of designing, launching and running a new business, which is often initially a small business. The people who create these businesses are called entrepreneurs.

Exam Probability: **Low**

29. *Answer choices:*

(see index for correct answer)

- a. Entrepreneurship
- b. Hayward
- c. Female entrepreneur
- d. Pit manager

Guidance: level 1

:: Management ::

A _____ is an idea of the future or desired result that a person or a group of people envisions, plans and commits to achieve. People endeavor to reach _____ s within a finite time by setting deadlines.

Exam Probability: **High**

30. *Answer choices:*

(see index for correct answer)

- a. Responsible autonomy
- b. Stakeholder
- c. Community-based management
- d. Dominant design

Guidance: level 1

:: Foreign direct investment ::

A _____ is an investment in the form of a controlling ownership in a business in one country by an entity based in another country. It is thus distinguished from a foreign portfolio investment by a notion of direct control.

Exam Probability: **Low**

31. *Answer choices:*

(see index for correct answer)

- a. Oligopolistic reaction
- b. Foreign direct investments in Kosovo
- c. Foreign direct investment
- d. Expropriation

Guidance: level 1

:: Accounting software ::

_____ is any item or verifiable record that is generally accepted as payment for goods and services and repayment of debts, such as taxes, in a particular country or socio-economic context. The main functions of _____ are distinguished as: a medium of exchange, a unit of account, a store of value and sometimes, a standard of deferred payment. Any item or verifiable record that fulfils these functions can be considered as _____ .

Exam Probability: **Medium**

32. *Answer choices:*

(see index for correct answer)

- a. DHPOS
- b. AME Accounting Software
- c. NewViews
- d. Money

Guidance: level 1

:: Stock market ::

A _____, securities exchange or bourse, is a facility where stock brokers and traders can buy and sell securities, such as shares of stock and bonds and other financial instruments. _____ s may also provide for facilities the issue and redemption of such securities and instruments and capital events including the payment of income and dividends. Securities traded on a _____ include stock issued by listed companies, unit trusts, derivatives, pooled investment products and bonds. _____ s often function as "continuous auction" markets with buyers and sellers consummating transactions via open outcry at a central location such as the floor of the exchange or by using an electronic trading platform.

Exam Probability: **High**

33. *Answer choices:*

(see index for correct answer)

- a. Stock exchange
- b. Security
- c. Cross listing
- d. Order book

Guidance: level 1

:: Private equity ::

_____ is a type of private equity, a form of financing that is provided by firms or funds to small, early-stage, emerging firms that are deemed to have high growth potential, or which have demonstrated high growth. _____ firms or funds invest in these early-stage companies in exchange for equity, or an ownership stake, in the companies they invest in. _____ists take on the risk of financing risky start-ups in the hopes that some of the firms they support will become successful. Because startups face high uncertainty, VC investments do have high rates of failure. The start-ups are usually based on an innovative technology or business model and they are usually from the high technology industries, such as information technology, clean technology or biotechnology.

Exam Probability: **Low**

34. *Answer choices:*

(see index for correct answer)

- a. Business Development Company
- b. Divisional buyout
- c. Firstpex
- d. Airwide Solutions

Guidance: level 1

:: Supply chain management ::

_____ is the process of finding and agreeing to terms, and acquiring goods, services, or works from an external source, often via a tendering or competitive bidding process. _____ is used to ensure the buyer receives goods, services, or works at the best possible price when aspects such as quality, quantity, time, and location are compared. Corporations and public bodies often define processes intended to promote fair and open competition for their business while minimizing risks such as exposure to fraud and collusion.

Exam Probability: **Low**

35. *Answer choices:*
(see index for correct answer)

- a. Dealer Business System
- b. Helveta
- c. Supply chain management software
- d. Procurement

Guidance: level 1

:: Business ::

_____ is a trade policy that does not restrict imports or exports; it can also be understood as the free market idea applied to international trade. In government, _____ is predominantly advocated by political parties that hold liberal economic positions while economically left-wing and nationalist political parties generally support protectionism, the opposite of _____.

Exam Probability: **Medium**

36. *Answer choices:*

(see index for correct answer)

- a. Business interaction networks
- b. Uncorporation
- c. Professional services
- d. Free trade

Guidance: level 1

:: Management ::

_____ is a process by which entities review the quality of all factors involved in production. ISO 9000 defines _____ as "A part of quality management focused on fulfilling quality requirements".

Exam Probability: **Medium**

37. *Answer choices:*

(see index for correct answer)

- a. Quality control
- b. Meeting
- c. Strategic management
- d. Context analysis

Guidance: level 1

:: Management ::

The _____ is a strategy performance management tool – a semi-standard structured report, that can be used by managers to keep track of the execution of activities by the staff within their control and to monitor the consequences arising from these actions.

Exam Probability: **Low**

38. *Answer choices:*

(see index for correct answer)

- a. Vendor relationship management
- b. Preparation
- c. Participative decision-making
- d. Balanced scorecard

Guidance: level 1

:: Critical thinking ::

In psychology, _____ is regarded as the cognitive process resulting in the selection of a belief or a course of action among several alternative possibilities. Every _____ process produces a final choice, which may or may not prompt action.

Exam Probability: **High**

39. *Answer choices:*

(see index for correct answer)

- a. Weak mindedness
- b. Project Reason
- c. Decision-making
- d. Straight face test

Guidance: level 1

:: Globalization-related theories ::

_____ is the process in which a nation is being improved in the sector of the economic, political, and social well-being of its people. The term has been used frequently by economists, politicians, and others in the 20th and 21st centuries. The concept, however, has been in existence in the West for centuries. "Modernization, "westernization", and especially "industrialization" are other terms often used while discussing _____. _____ has a direct relationship with the environment and environmental issues. _____ is very often confused with industrial development, even in some academic sources.

Exam Probability: **High**

40. *Answer choices:*

(see index for correct answer)

- a. post-industrial
- b. postmodernism
- c. Economic Development

Guidance: level 1

:: Alchemical processes ::

In chemistry, a _____ is a special type of homogeneous mixture composed of two or more substances. In such a mixture, a solute is a substance dissolved in another substance, known as a solvent. The mixing process of a _____ happens at a scale where the effects of chemical polarity are involved, resulting in interactions that are specific to solvation. The _____ assumes the phase of the solvent when the solvent is the larger fraction of the mixture, as is commonly the case. The concentration of a solute in a _____ is the mass of that solute expressed as a percentage of the mass of the whole _____ . The term aqueous _____ is when one of the solvents is water.

Exam Probability: **Low**

41. *Answer choices:*

(see index for correct answer)

- a. Unity of opposites

- b. Fermentation in food processing
- c. Solution
- d. Fixation

Guidance: level 1

:: Management ::

_____ is the practice of initiating, planning, executing, controlling, and closing the work of a team to achieve specific goals and meet specific success criteria at the specified time.

Exam Probability: **Medium**

42. *Answer choices:*

(see index for correct answer)

- a. Main Street Manager
- b. Quality control
- c. Board of governors
- d. Project management

Guidance: level 1

:: Marketing ::

A _____ is a group of customers within a business's serviceable available market at which a business aims its marketing efforts and resources. A _____ is a subset of the total market for a product or service. The _____ typically consists of consumers who exhibit similar characteristics and are considered most likely to buy a business's market offerings or are likely to be the most profitable segments for the business to service.

Exam Probability: **Medium**

43. *Answer choices:*

(see index for correct answer)

- a. Product planning
- b. Target market
- c. Existing visitor optimisation
- d. Hakan Okay

Guidance: level 1

:: Cash flow ::

_____ s are narrowly interconnected with the concepts of value, interest rate and liquidity. A _____ that shall happen on a future day tN can be transformed into a _____ of the same value in t0.

Exam Probability: **Medium**

44. Answer choices:

(see index for correct answer)

- a. Valuation using discounted cash flows
- b. Factoring
- c. Cash flow
- d. Cash flow forecasting

Guidance: level 1

:: International trade ::

An _____ is a good brought into a jurisdiction, especially across a national border, from an external source. The party bringing in the good is called an _____ er. An _____ in the receiving country is an export from the sending country. _____ ation and exportation are the defining financial transactions of international trade.

Exam Probability: **Low**

45. Answer choices:

(see index for correct answer)

- a. Trade in services
- b. International Association for Technology Trade
- c. Import
- d. Debt moratorium

Guidance: level 1

:: Office administration ::

An _____ is generally a room or other area where an organization's employees perform administrative work in order to support and realize objects and goals of the organization. The word "_____" may also denote a position within an organization with specific duties attached to it ; the latter is in fact an earlier usage, _____ as place originally referring to the location of one's duty. When used as an adjective, the term "_____" may refer to business-related tasks. In law, a company or organization has _____ s in any place where it has an official presence, even if that presence consists of a storage silo rather than an establishment with desk-and-chair. An _____ is also an architectural and design phenomenon: ranging from a small _____ such as a bench in the corner of a small business of extremely small size , through entire floors of buildings, up to and including massive buildings dedicated entirely to one company. In modern terms an _____ is usually the location where white-collar workers carry out their functions. As per James Stephenson, "_____ is that part of business enterprise which is devoted to the direction and co-ordination of its various activities."

Exam Probability: **Low**

46. *Answer choices:*

(see index for correct answer)

- a. Office administration
- b. Inter departmental communication
- c. Fish! Philosophy
- d. Office

Guidance: level 1

:: Project management ::

In political science, an _____ is a means by which a petition signed by a certain minimum number of registered voters can force a government to choose to either enact a law or hold a public vote in parliament in what is called indirect _____ , or under direct _____ , the proposition is immediately put to a plebiscite or referendum, in what is called a Popular initiated Referendum or citizen-initiated referendum).

Exam Probability: **Medium**

47. *Answer choices:*

(see index for correct answer)

- a. Scope statement
- b. Concept note
- c. Rolling Wave planning
- d. Extreme project management

Guidance: level 1

:: Derivatives (finance) ::

_____ is any bodily activity that enhances or maintains physical fitness and overall health and wellness. It is performed for various reasons, to aid growth and improve strength, preventing aging, developing muscles and the cardiovascular system, honing athletic skills, weight loss or maintenance, improving health and also for enjoyment. Many individuals choose to _____ outdoors where they can congregate in groups, socialize, and enhance well-being.

Exam Probability: **Medium**

48. *Answer choices:*

(see index for correct answer)

- a. Commodity tick
- b. Dollar roll
- c. Exercise
- d. ISDA Master Agreement

Guidance: level 1

:: ::

Some scenarios associate "this kind of planning" with learning "life skills". Schedules are necessary, or at least useful, in situations where individuals need to know what time they must be at a specific location to receive a specific service, and where people need to accomplish a set of goals within a set time period.

Exam Probability: **Low**

49. *Answer choices:*

(see index for correct answer)

- a. corporate values
- b. Character
- c. information systems assessment
- d. imperative

Guidance: level 1

:: Casting (manufacturing) ::

A _____ is a regularity in the world, man-made design, or abstract ideas. As such, the elements of a _____ repeat in a predictable manner. A geometric _____ is a kind of _____ formed of geometric shapes and typically repeated like a wallpaper design.

Exam Probability: **High**

50. *Answer choices:*

(see index for correct answer)

- a. AutoCAST
- b. Pattern
- c. Castability

- d. Bronze sculpture

Guidance: level 1

:: Mereology ::

_____ , in the abstract, is what belongs to or with something, whether as an attribute or as a component of said thing. In the context of this article, it is one or more components , whether physical or incorporeal, of a person's estate; or so belonging to, as in being owned by, a person or jointly a group of people or a legal entity like a corporation or even a society. Depending on the nature of the _____ , an owner of _____ has the right to consume, alter, share, redefine, rent, mortgage, pawn, sell, exchange, transfer, give away or destroy it, or to exclude others from doing these things, as well as to perhaps abandon it; whereas regardless of the nature of the _____ , the owner thereof has the right to properly use it , or at the very least exclusively keep it.

Exam Probability: **High**

51. *Answer choices:*
(see index for correct answer)

- a. Non-wellfounded mereology
- b. Property
- c. Simple
- d. Gunk

Guidance: level 1

:: Organizational structure ::

An _____ defines how activities such as task allocation, coordination, and supervision are directed toward the achievement of organizational aims.

Exam Probability: **Low**

52. *Answer choices:*

(see index for correct answer)

- a. Followership
- b. Automated Bureaucracy
- c. Organization of the New York City Police Department
- d. Organizational structure

Guidance: level 1

:: Basic financial concepts ::

_____ is a sustained increase in the general price level of goods and services in an economy over a period of time. When the general price level rises, each unit of currency buys fewer goods and services; consequently, _____ reflects a reduction in the purchasing power per unit of money a loss of real value in the medium of exchange and unit of account within the economy. The measure of _____ is the _____ rate, the annualized percentage change in a general price index, usually the consumer price index, over time. The opposite of _____ is deflation.

Exam Probability: **Low**

53. *Answer choices:*

(see index for correct answer)

- a. Financial transaction
- b. Leverage cycle
- c. Forward guidance
- d. Inflation

Guidance: level 1

:: Interest rates ::

An _____ is the amount of interest due per period, as a proportion of the amount lent, deposited or borrowed. The total interest on an amount lent or borrowed depends on the principal sum, the _____ , the compounding frequency, and the length of time over which it is lent, deposited or borrowed.

Exam Probability: **Medium**

54. *Answer choices:*

(see index for correct answer)

- a. Forex swap
- b. Rate
- c. Interest rate
- d. Foreign exchange swap

Guidance: level 1

:: Fraud ::

In law, _____ is intentional deception to secure unfair or unlawful gain, or to deprive a victim of a legal right. _____ can violate civil law, a criminal law, or it may cause no loss of money, property or legal right but still be an element of another civil or criminal wrong. The purpose of _____ may be monetary gain or other benefits, for example by obtaining a passport, travel document, or driver's license, or mortgage _____, where the perpetrator may attempt to qualify for a mortgage by way of false statements.

Exam Probability: **Low**

55. *Answer choices:*

(see index for correct answer)

- a. Check kiting
- b. Welfare queen
- c. Lip-synching in music
- d. Fraud

Guidance: level 1

:: ::

_____ or accountancy is the measurement, processing, and communication of financial information about economic entities such as businesses and corporations. The modern field was established by the Italian mathematician Luca Pacioli in 1494. _____ , which has been called the "language of business", measures the results of an organization's economic activities and conveys this information to a variety of users, including investors, creditors, management, and regulators. Practitioners of _____ are known as accountants. The terms " _____ " and "financial reporting" are often used as synonyms.

Exam Probability: **High**

56. *Answer choices:*

(see index for correct answer)

- a. Accounting
- b. hierarchical perspective
- c. surface-level diversity
- d. open system

Guidance: level 1

:: Industrial design ::

In physics and mathematics, the _____ of a mathematical space is informally defined as the minimum number of coordinates needed to specify any point within it. Thus a line has a _____ of one because only one coordinate is needed to specify a point on it for example, the point at 5 on a number line. A surface such as a plane or the surface of a cylinder or sphere has a _____ of two because two coordinates are needed to specify a point on it for example, both a latitude and longitude are required to locate a point on the surface of a sphere. The inside of a cube, a cylinder or a sphere is three- _____ al because three coordinates are needed to locate a point within these spaces.

Exam Probability: **Medium**

57. *Answer choices:*
(see index for correct answer)

- a. Dimension
- b. Sustainable furniture design
- c. Design for X
- d. ModeMapping

Guidance: level 1

:: Financial markets ::

A _____ is a financial market in which long-term debt or equity-backed securities are bought and sold. _____ s channel the wealth of savers to those who can put it to long-term productive use, such as companies or governments making long-term investments. Financial regulators like the Bank of England and the U.S. Securities and Exchange Commission oversee _____ s to protect investors against fraud, among other duties.

Exam Probability: **Low**

58. *Answer choices:*

(see index for correct answer)

- a. Pfandbrief
- b. Flight-to-liquidity
- c. Secondary market
- d. Broker-dealer

Guidance: level 1

:: ::

A _____ is any person who contracts to acquire an asset in return for some form of consideration.

Exam Probability: **Medium**

59. *Answer choices:*

(see index for correct answer)

- a. open system
- b. interpersonal communication
- c. Character
- d. Buyer

Guidance: level 1

Management

Management is the administration of an organization, whether it is a business, a not-for-profit organization, or government body. Management includes the activities of setting the strategy of an organization and coordinating the efforts of its employees (or of volunteers) to accomplish its objectives through the application of available resources, such as financial, natural, technological, and human resources.

:: Hospitality management ::

A _____ is an establishment that provides paid lodging on a short-term basis. Facilities provided may range from a modest-quality mattress in a small room to large suites with bigger, higher-quality beds, a dresser, a refrigerator and other kitchen facilities, upholstered chairs, a flat screen television, and en-suite bathrooms. Small, lower-priced _____ s may offer only the most basic guest services and facilities. Larger, higher-priced _____ s may provide additional guest facilities such as a swimming pool, business centre, childcare, conference and event facilities, tennis or basketball courts, gymnasium, restaurants, day spa, and social function services. _____ rooms are usually numbered to allow guests to identify their room. Some boutique, high-end _____ s have custom decorated rooms. Some _____ s offer meals as part of a room and board arrangement. In the United Kingdom, a _____ is required by law to serve food and drinks to all guests within certain stated hours. In Japan, capsule _____ s provide a tiny room suitable only for sleeping and shared bathroom facilities.

Exam Probability: **High**

1. *Answer choices:*

(see index for correct answer)

- a. Birgit Zotz
- b. Hotel
- c. Towel animal
- d. City ledger

Guidance: level 1

:: Management ::

A _____ is when two or more people come together to discuss one or more topics, often in a formal or business setting, but _____ s also occur in a variety of other environments. Many various types of _____ s exist.

Exam Probability: **Low**

2. *Answer choices:*

(see index for correct answer)

- a. Design management
- b. Planning
- c. U-procedure and Theory U
- d. middle manager

Guidance: level 1

:: ::

A _____ , or also known as foreman, overseer, facilitator, monitor, area coordinator, or sometimes gaffer, is the job title of a low level management position that is primarily based on authority over a worker or charge of a workplace. A _____ can also be one of the most senior in the staff at the place of work, such as a Professor who oversees a PhD dissertation. Supervision, on the other hand, can be performed by people without this formal title, for example by parents. The term _____ itself can be used to refer to any personnel who have this task as part of their job description.

Exam Probability: **Low**

3. Answer choices:

(see index for correct answer)

- a. information systems assessment
- b. cultural
- c. corporate values
- d. functional perspective

Guidance: level 1

:: Business ::

_____ is a trade policy that does not restrict imports or exports; it can also be understood as the free market idea applied to international trade. In government, _____ is predominantly advocated by political parties that hold liberal economic positions while economically left-wing and nationalist political parties generally support protectionism, the opposite of _____ .

Exam Probability: **Medium**

4. Answer choices:

(see index for correct answer)

- a. Business history
- b. Free trade
- c. Policy capturing
- d. 24/7 service

Guidance: level 1

:: Evaluation ::

A _____ is an evaluation of a publication, service, or company such as a movie, video game, musical composition, book; a piece of hardware like a car, home appliance, or computer; or an event or performance, such as a live music concert, play, musical theater show, dance show, or art exhibition. In addition to a critical evaluation, the _____'s author may assign the work a rating to indicate its relative merit. More loosely, an author may _____ current events, trends, or items in the news. A compilation of _____ s may itself be called a _____. The New York _____ of Books, for instance, is a collection of essays on literature, culture, and current affairs. National _____, founded by William F. Buckley, Jr., is an influential conservative magazine, and Monthly _____ is a long-running socialist periodical.

Exam Probability: **Low**

5. *Answer choices:*
(see index for correct answer)

- a. Quality assurance
- b. Impact assessment
- c. Continuous assessment
- d. Princeton Application Repository for Shared-Memory Computers

Guidance: level 1

:: Business models ::

_____ es are privately owned corporations, partnerships, or sole proprietorships that have fewer employees and/or less annual revenue than a regular-sized business or corporation. Businesses are defined as "small" in terms of being able to apply for government support and qualify for preferential tax policy varies depending on the country and industry. _____ es range from fifteen employees under the Australian Fair Work Act 2009, fifty employees according to the definition used by the European Union, and fewer than five hundred employees to qualify for many U.S. _____ Administration programs. While _____ es can also be classified according to other methods, such as annual revenues, shipments, sales, assets, or by annual gross or net revenue or net profits, the number of employees is one of the most widely used measures.

Exam Probability: **Medium**

6. *Answer choices:*

(see index for correct answer)

- a. The Community Company
- b. Open Music Model
- c. Component business model
- d. Small business

Guidance: level 1

:: Evaluation ::

_____ is the practice of being honest and showing a consistent and uncompromising adherence to strong moral and ethical principles and values. In ethics, _____ is regarded as the honesty and truthfulness or accuracy of one's actions. _____ can stand in opposition to hypocrisy, in that judging with the standards of _____ involves regarding internal consistency as a virtue, and suggests that parties holding within themselves apparently conflicting values should account for the discrepancy or alter their beliefs. The word _____ evolved from the Latin adjective integer, meaning whole or complete. In this context, _____ is the inner sense of "wholeness" deriving from qualities such as honesty and consistency of character. As such, one may judge that others "have _____" to the extent that they act according to the values, beliefs and principles they claim to hold.

Exam Probability: **Medium**

7. *Answer choices:*

(see index for correct answer)

- a. Australian Drug Evaluation Committee
- b. Integrity
- c. Formative assessment
- d. Academic equivalency evaluation

Guidance: level 1

:: Employment discrimination ::

A _____ is a metaphor used to represent an invisible barrier that keeps a given demographic from rising beyond a certain level in a hierarchy.

Exam Probability: **Medium**

8. *Answer choices:*

(see index for correct answer)

- a. Employment discrimination law in the European Union
- b. MacBride Principles
- c. Marriage bars
- d. LGBT employment discrimination in the United States

Guidance: level 1

:: Business models ::

A _____ is "an autonomous association of persons united voluntarily to meet their common economic, social, and cultural needs and aspirations through a jointly-owned and democratically-controlled enterprise". _____ s may include.

Exam Probability: **High**

9. *Answer choices:*

(see index for correct answer)

- a. Co-operative economics
- b. Cooperative
- c. Micro-enterprise
- d. What if chart

Guidance: level 1

:: Management ::

The term _____ refers to measures designed to increase the degree of autonomy and self-determination in people and in communities in order to enable them to represent their interests in a responsible and self-determined way, acting on their own authority. It is the process of becoming stronger and more confident, especially in controlling one's life and claiming one's rights. _____ as action refers both to the process of self-_____ and to professional support of people, which enables them to overcome their sense of powerlessness and lack of influence, and to recognize and use their resources. To do work with power.

Exam Probability: **Medium**

10. *Answer choices:*

(see index for correct answer)

- a. Social business model
- b. Infrastructure asset management
- c. Value migration
- d. Peer pressure

Guidance: level 1

:: Management occupations ::

_____ is the process of designing, launching and running a new business, which is often initially a small business. The people who create these businesses are called entrepreneurs.

Exam Probability: **Low**

11. *Answer choices:*

(see index for correct answer)

- a. Chief gaming officer
- b. Functional manager
- c. Chief design officer
- d. Entrepreneurship

Guidance: level 1

:: Decision theory ::

Within economics the concept of _____ is used to model worth or value, but its usage has evolved significantly over time. The term was introduced initially as a measure of pleasure or satisfaction within the theory of utilitarianism by moral philosophers such as Jeremy Bentham and John Stuart Mill. But the term has been adapted and reapplied within neoclassical economics, which dominates modern economic theory, as a _____ function that represents a consumer's preference ordering over a choice set. As such, it is devoid of its original interpretation as a measurement of the pleasure or satisfaction obtained by the consumer from that choice.

Exam Probability: **Medium**

12. *Answer choices:*

(see index for correct answer)

- a. Utility
- b. Regret
- c. Social and Decision Sciences
- d. Belief decision matrix

Guidance: level 1

:: ::

_____ is the amount of time someone works beyond normal working hours. The term is also used for the pay received for this time. Normal hours may be determined in several ways.

Exam Probability: **Medium**

13. *Answer choices:*

(see index for correct answer)

- a. Overtime
- b. personal values
- c. deep-level diversity
- d. levels of analysis

Guidance: level 1

:: ::

_____ is an evaluative or corrective exercise that can occur in any area of human life. _____ can therefore take many different forms. How people go about criticizing, can vary a great deal. In specific areas of human endeavour, the form of _____ can be highly specialized and technical; it often requires professional knowledge to appreciate the _____. For subject-specific information, see the Varieties of _____ page.

Exam Probability: **Low**

14. *Answer choices:*

(see index for correct answer)

- a. functional perspective
- b. hierarchical perspective

- c. open system
- d. similarity-attraction theory

Guidance: level 1

:: Management accounting ::

_____ s are costs that change as the quantity of the good or service that a business produces changes. _____ s are the sum of marginal costs over all units produced. They can also be considered normal costs. Fixed costs and _____ s make up the two components of total cost. Direct costs are costs that can easily be associated with a particular cost object. However, not all _____ s are direct costs. For example, variable manufacturing overhead costs are _____ s that are indirect costs, not direct costs. _____ s are sometimes called unit-level costs as they vary with the number of units produced.

Exam Probability: **High**

15. *Answer choices:*

(see index for correct answer)

- a. Institute of Cost and Management Accountants of Bangladesh
- b. activity based costing
- c. Variable cost
- d. Fixed assets management

Guidance: level 1

:: ::

An _____ in international trade is a good or service produced in one country that is bought by someone in another country. The seller of such goods and services is an _____ er; the foreign buyer is an importer.

Exam Probability: **High**

16. *Answer choices:*

(see index for correct answer)

- a. empathy
- b. Sarbanes-Oxley act of 2002
- c. deep-level diversity
- d. information systems assessment

Guidance: level 1

:: ::

In logic and philosophy, an _____ is a series of statements, called the premises or premisses, intended to determine the degree of truth of another statement, the conclusion. The logical form of an _____ in a natural language can be represented in a symbolic formal language, and independently of natural language formally defined "_____ s" can be made in math and computer science.

Exam Probability: **Low**

17. *Answer choices:*

(see index for correct answer)

- a. co-culture
- b. open system
- c. Character
- d. Argument

Guidance: level 1

:: ::

_____, in its broadest context, includes both the attainment of that which is just and the philosophical discussion of that which is just. The concept of _____ is based on numerous fields, and many differing viewpoints and perspectives including the concepts of moral correctness based on ethics, rationality, law, religion, equity and fairness. Often, the general discussion of _____ is divided into the realm of social _____ as found in philosophy, theology and religion, and, procedural _____ as found in the study and application of the law.

Exam Probability: **High**

18. *Answer choices:*

(see index for correct answer)

- a. functional perspective
- b. imperative
- c. information systems assessment
- d. similarity-attraction theory

Guidance: level 1

:: Labour relations ::

_____ is a field of study that can have different meanings depending on the context in which it is used. In an international context, it is a subfield of labor history that studies the human relations with regard to work – in its broadest sense – and how this connects to questions of social inequality. It explicitly encompasses unregulated, historical, and non-Western forms of labor. Here, _____ define "for or with whom one works and under what rules. These rules determine the type of work, type and amount of remuneration, working hours, degrees of physical and psychological strain, as well as the degree of freedom and autonomy associated with the work."

Exam Probability: **High**

19. *Answer choices:*
(see index for correct answer)

- a. United Students Against Sweatshops
- b. Work Order Act
- c. Inflatable rat
- d. Labor relations

Guidance: level 1

:: Budgets ::

A _____ is a financial plan for a defined period, often one year. It may also include planned sales volumes and revenues, resource quantities, costs and expenses, assets, liabilities and cash flows. Companies, governments, families and other organizations use it to express strategic plans of activities or events in measurable terms.

Exam Probability: **High**

20. *Answer choices:*

(see index for correct answer)

- a. Zero budget
- b. Marginal budgeting for bottlenecks
- c. Participatory budgeting
- d. Budgeted cost of work scheduled

Guidance: level 1

:: Management occupations ::

_____ship is the process of designing, launching and running a new business, which is often initially a small business. The people who create these businesses are called _____s.

Exam Probability: **Low**

21. *Answer choices:*

(see index for correct answer)

- a. Corporate trainer
- b. Business magnate
- c. Female entrepreneur
- d. Chief business development officer

Guidance: level 1

:: Business ::

The seller, or the provider of the goods or services, completes a sale in response to an acquisition, appropriation, requisition or a direct interaction with the buyer at the point of sale. There is a passing of title of the item, and the settlement of a price, in which agreement is reached on a price for which transfer of ownership of the item will occur. The seller, not the purchaser typically executes the sale and it may be completed prior to the obligation of payment. In the case of indirect interaction, a person who sells goods or service on behalf of the owner is known as a _____ man or _____ woman or _____ person, but this often refers to someone selling goods in a store/shop, in which case other terms are also common, including _____ clerk, shop assistant, and retail clerk.

Exam Probability: **High**

22. *Answer choices:*

(see index for correct answer)

- a. Religion and business
- b. Corporate housing
- c. Backward invention
- d. Ametek

Guidance: level 1

:: Occupational safety and health ::

_____ is a chemical element with symbol Pb and atomic number 82. It is a heavy metal that is denser than most common materials. _____ is soft and malleable, and also has a relatively low melting point. When freshly cut, _____ is silvery with a hint of blue; it tarnishes to a dull gray color when exposed to air. _____ has the highest atomic number of any stable element and three of its isotopes are endpoints of major nuclear decay chains of heavier elements.

Exam Probability: **Medium**

23. *Answer choices:*

(see index for correct answer)

- a. Formaldehyde

- b. Nutec
- c. Manganese
- d. Lead

Guidance: level 1

:: Problem solving ::

A _____ is a unit or formation established to work on a single defined task or activity. Originally introduced by the United States Navy, the term has now caught on for general usage and is a standard part of NATO terminology. Many non-military organizations now create " _____ s" or task groups for temporary activities that might have once been performed by ad hoc committees.

Exam Probability: **Medium**

24. *Answer choices:*
(see index for correct answer)

- a. How to Solve It
- b. Task force
- c. Problem finding
- d. Trizics

Guidance: level 1

:: Organizational theory ::

A _____ is an organizational theory that claims that there is no best way to organize a corporation, to lead a company, or to make decisions. Instead, the optimal course of action is contingent upon the internal and external situation. A contingent leader effectively applies their own style of leadership to the right situation.

Exam Probability: **Medium**

25. *Answer choices:*
(see index for correct answer)

- a. Seagull manager
- b. Organisational semiotics
- c. Organization theory
- d. Organizational effectiveness

Guidance: level 1

:: Reputation management ::

_____ or image of a social entity is an opinion about that entity, typically as a result of social evaluation on a set of criteria.

Exam Probability: **Low**

26. Answer choices:

(see index for correct answer)

- a. Advogato
- b. Lithium Technologies
- c. Reputation
- d. Star

Guidance: level 1

:: Personality tests ::

The Myers–Briggs Type Indicator is an introspective self-report questionnaire with the purpose of indicating differing psychological preferences in how people perceive the world around them and make decisions. . Though the test superficially resembles some psychological theories it is commonly classified as pseudoscience, especially as pertains to its supposed predictive abilities.

Exam Probability: **Low**

27. Answer choices:

(see index for correct answer)

- a. Myers-Briggs type
- b. Keirsey Temperament Sorter
- c. Myers-Briggs Type Indicator
- d. personality quiz

Guidance: level 1

:: Leadership ::

> _____ Theory, or the _____ Model, is a model created by Paul Hersey and Ken Blanchard, developed while working on Management of Organizational Behavior. The theory was first introduced in 1969 as "life cycle theory of leadership". During the mid-1970s, life cycle theory of leadership was renamed " _____ Theory."

Exam Probability: **High**

28. *Answer choices:*

(see index for correct answer)

- a. Love leadership
- b. Situational leadership
- c. Maintenance actions
- d. Moral example

Guidance: level 1

:: ::

In production, research, retail, and accounting, a _____ is the value of money that has been used up to produce something or deliver a service, and hence is not available for use anymore. In business, the _____ may be one of acquisition, in which case the amount of money expended to acquire it is counted as _____ . In this case, money is the input that is gone in order to acquire the thing. This acquisition _____ may be the sum of the _____ of production as incurred by the original producer, and further _____ s of transaction as incurred by the acquirer over and above the price paid to the producer. Usually, the price also includes a mark-up for profit over the _____ of production.

Exam Probability: **Low**

29. *Answer choices:*

(see index for correct answer)

- a. imperative
- b. hierarchical perspective
- c. co-culture
- d. levels of analysis

Guidance: level 1

_____ is the consumption and saving opportunity gained by an entity within a specified timeframe, which is generally expressed in monetary terms. For households and individuals, " _____ is the sum of all the wages, salaries, profits, interest payments, rents, and other forms of earnings received in a given period of time."

Exam Probability: **High**

30. *Answer choices:*

(see index for correct answer)

- a. corporate values
- b. deep-level diversity
- c. similarity-attraction theory
- d. Income

Guidance: level 1

:: Marketing techniques ::

In industry, product lifecycle management is the process of managing the entire lifecycle of a product from inception, through engineering design and manufacture, to service and disposal of manufactured products. PLM integrates people, data, processes and business systems and provides a product information backbone for companies and their extended enterprise.

Exam Probability: **High**

31. *Answer choices:*

(see index for correct answer)

- a. SONCAS
- b. Relevant space
- c. Loss leader
- d. Product life cycle

Guidance: level 1

:: Grounds for termination of employment ::

_____ is a habitual pattern of absence from a duty or obligation without good reason. Generally, _____ is unplanned absences. _____ has been viewed as an indicator of poor individual performance, as well as a breach of an implicit contract between employee and employer. It is seen as a management problem, and framed in economic or quasi-economic terms. More recent scholarship seeks to understand _____ as an indicator of psychological, medical, or social adjustment to work.

Exam Probability: **High**

32. *Answer choices:*

(see index for correct answer)

- a. Department of Defense Whistleblower Program
- b. Defense Intelligence Community Whistleblower Protection
- c. Huffman v. Office of Personnel Management

- d. No call, no show

Guidance: level 1

:: Product management ::

_____ s, also known as Shewhart charts or process-behavior charts, are a statistical process control tool used to determine if a manufacturing or business process is in a state of control.

Exam Probability: **Medium**

33. *Answer choices:*
(see index for correct answer)

- a. Diffusion of innovations
- b. Tipping point
- c. Product information
- d. Control chart

Guidance: level 1

:: Production economics ::

_____ is the creation of a whole that is greater than the simple sum of its parts. The term _____ comes from the Attic Greek word sea synergia from synergos, , meaning "working together".

Exam Probability: **High**

34. *Answer choices:*
(see index for correct answer)

- a. Industrial production index
- b. Marginal product
- c. Isoquant
- d. Post-Fordism

Guidance: level 1

:: Poker strategy ::

_____ is any measure taken to guard a thing against damage caused by outside forces. _____ can be provided to physical objects, including organisms, to systems, and to intangible things like civil and political rights. Although the mechanisms for providing _____ vary widely, the basic meaning of the term remains the same. This is illustrated by an explanation found in a manual on electrical wiring.

Exam Probability: **Medium**

35. *Answer choices:*

(see index for correct answer)

- a. Position
- b. Fundamental theorem of poker
- c. Protection
- d. Slow play

Guidance: level 1

:: Project management ::

_____ is a process of setting goals, planning and/or controlling the organizing and leading the execution of any type of activity, such as.

Exam Probability: **Low**

36. *Answer choices:*

(see index for correct answer)

- a. Level of Effort
- b. Management process
- c. Axelos
- d. Project governance

Guidance: level 1

:: Operations research ::

_____ is a method to achieve the best outcome in a mathematical model whose requirements are represented by linear relationships. _____ is a special case of mathematical programming .

Exam Probability: **Medium**

37. *Answer choices:*

(see index for correct answer)

- a. Inventory theory
- b. DYNAMO
- c. Linear programming
- d. Canadian traveller problem

Guidance: level 1

:: Marketing ::

_____ or stock control can be broadly defined as "the activity of checking a shop's stock." However, a more focused definition takes into account the more science-based, methodical practice of not only verifying a business' inventory but also focusing on the many related facets of inventory management "within an organisation to meet the demand placed upon that business economically." Other facets of _____ include supply chain management, production control, financial flexibility, and customer satisfaction. At the root of _____, however, is the _____ problem, which involves determining when to order, how much to order, and the logistics of those decisions.

Exam Probability: **Low**

38. *Answer choices:*

(see index for correct answer)

- a. Postmodern communication
- b. Branding national myths and symbols
- c. Inventory control
- d. Impulse buying

Guidance: level 1

:: Evaluation ::

_____ solving consists of using generic or ad hoc methods in an orderly manner to find solutions to _____ s. Some of the _____ -solving techniques developed and used in philosophy, artificial intelligence, computer science, engineering, mathematics, or medicine are related to mental _____ -solving techniques studied in psychology.

Exam Probability: **Low**

39. *Answer choices:*

(see index for correct answer)

- a. Health technology assessment
- b. Technology assessment
- c. Problem
- d. Joint Committee on Standards for Educational Evaluation

Guidance: level 1

:: Cognitive biases ::

The _____ is a type of immediate judgement discrepancy, or cognitive bias, where a person making an initial assessment of another person, place, or thing will assume ambiguous information based upon concrete information. A simplified example of the _____ is when an individual noticing that the person in the photograph is attractive, well groomed, and properly attired, assumes, using a mental heuristic, that the person in the photograph is a good person based upon the rules of that individual's social concept. This constant error in judgment is reflective of the individual's preferences, prejudices, ideology, aspirations, and social perception. The _____ is an evaluation by an individual and can affect the perception of a decision, action, idea, business, person, group, entity, or other whenever concrete data is generalized or influences ambiguous information.

Exam Probability: **Low**

40. *Answer choices:*

(see index for correct answer)

- a. Forer effect
- b. Picture superiority effect
- c. Just-world hypothesis
- d. Halo effect

Guidance: level 1

:: Production and manufacturing ::

Automatic _____ in continuous production processes is a combination of control engineering and chemical engineering disciplines that uses industrial control systems to achieve a production level of consistency, economy and safety which could not be achieved purely by human manual control. It is implemented widely in industries such as oil refining, pulp and paper manufacturing, chemical processing and power generating plants.

Exam Probability: **Medium**

41. *Answer choices:*

(see index for correct answer)

- a. First pass yield
- b. Hydrosila
- c. Process control
- d. Value engineering

Guidance: level 1

:: Summary statistics ::

_____ is the number of occurrences of a repeating event per unit of time. It is also referred to as temporal _____, which emphasizes the contrast to spatial _____ and angular _____. The period is the duration of time of one cycle in a repeating event, so the period is the reciprocal of the _____. For example: if a newborn baby's heart beats at a _____ of 120 times a minute, its period—the time interval between beats—is half a second. _____ is an important parameter used in science and engineering to specify the rate of oscillatory and vibratory phenomena, such as mechanical vibrations, audio signals, radio waves, and light.

Exam Probability: **Medium**

42. *Answer choices:*

(see index for correct answer)

- a. Frequency
- b. Scan statistic
- c. Multiple of the median
- d. Quantile

Guidance: level 1

:: Project management ::

In political science, an _____ is a means by which a petition signed by a certain minimum number of registered voters can force a government to choose to either enact a law or hold a public vote in parliament in what is called indirect _____ , or under direct _____ , the proposition is immediately put to a plebiscite or referendum, in what is called a Popular initiated Referendum or citizen-initiated referendum).

Exam Probability: **Medium**

43. *Answer choices:*

(see index for correct answer)

- a. Scope
- b. Initiative
- c. Value breakdown structure
- d. Changes clause

Guidance: level 1

:: Survey methodology ::

An _____ is a conversation where questions are asked and answers are given. In common parlance, the word " _____ " refers to a one-on-one conversation between an _____ er and an _____ ee. The _____ er asks questions to which the _____ ee responds, usually so information may be transferred from _____ ee to _____ er . Sometimes, information can be transferred in both directions. It is a communication, unlike a speech, which produces a one-way flow of information.

Exam Probability: **Medium**

44. *Answer choices:*

(see index for correct answer)

- a. Coverage error
- b. American Association for Public Opinion Research
- c. Survey research
- d. Group concept mapping

Guidance: level 1

:: ::

An _____ is the production of goods or related services within an economy. The major source of revenue of a group or company is the indicator of its relevant _____. When a large group has multiple sources of revenue generation, it is considered to be working in different industries. Manufacturing _____ became a key sector of production and labour in European and North American countries during the Industrial Revolution, upsetting previous mercantile and feudal economies. This came through many successive rapid advances in technology, such as the production of steel and coal.

Exam Probability: **Medium**

45. *Answer choices:*

(see index for correct answer)

- a. hierarchical
- b. Industry
- c. similarity-attraction theory
- d. levels of analysis

Guidance: level 1

:: Power (social and political) ::

_____ is a form of reverence gained by a leader who has strong interpersonal relationship skills. _____ , as an aspect of personal power, becomes particularly important as organizational leadership becomes increasingly about collaboration and influence, rather than command and control.

Exam Probability: **High**

46. *Answer choices:*
(see index for correct answer)

- a. Hard power
- b. Expert power
- c. Referent power

Guidance: level 1

:: Human resource management ::

_____ is the corporate management term for the act of reorganizing the legal, ownership, operational, or other structures of a company for the purpose of making it more profitable, or better organized for its present needs. Other reasons for _____ include a change of ownership or ownership structure, demerger, or a response to a crisis or major change in the business such as bankruptcy, repositioning, or buyout. _____ may also be described as corporate _____, debt _____ and financial _____.

Exam Probability: **Low**

47. *Answer choices:*

(see index for correct answer)

- a. On-ramping
- b. Virtual management
- c. Restructuring
- d. IDS HR in Practice

Guidance: level 1

:: ::

An _____ is a person temporarily or permanently residing in a country other than their native country. In common usage, the term often refers to professionals, skilled workers, or artists taking positions outside their home country, either independently or sent abroad by their employers, who can be companies, universities, governments, or non-governmental organisations. Effectively migrant workers, they usually earn more than they would at home, and less than local employees. However, the term ` _____ ` is also used for retirees and others who have chosen to live outside their native country. Historically, it has also referred to exiles.

Exam Probability: **Medium**

48. *Answer choices:*
(see index for correct answer)

- a. Character
- b. surface-level diversity
- c. deep-level diversity
- d. empathy

Guidance: level 1

:: Human resource management ::

_____ is a core function of human resource management and it is related to the specification of contents, methods and relationship of jobs in order to satisfy technological and organizational requirements as well as the social and personal requirements of the job holder or the employee. Its principles are geared towards how the nature of a person's job affects their attitudes and behavior at work, particularly relating to characteristics such as skill variety and autonomy. The aim of a _____ is to improve job satisfaction, to improve through-put, to improve quality and to reduce employee problems .

Exam Probability: **Low**

49. *Answer choices:*

(see index for correct answer)

- a. Management due diligence
- b. Job design
- c. TPI-theory
- d. Employee value proposition

Guidance: level 1

:: ::

_____ is a form of development in which a person called a coach supports a learner or client in achieving a specific personal or professional goal by providing training and guidance. The learner is sometimes called a coachee. Occasionally, _____ may mean an informal relationship between two people, of whom one has more experience and expertise than the other and offers advice and guidance as the latter learns; but _____ differs from mentoring in focusing on specific tasks or objectives, as opposed to more general goals or overall development.

Exam Probability: **Medium**

50. *Answer choices:*

(see index for correct answer)

- a. information systems assessment
- b. cultural
- c. personal values
- d. similarity-attraction theory

Guidance: level 1

:: Organizational theory ::

_____ is the process of creating, retaining, and transferring knowledge within an organization. An organization improves over time as it gains experience. From this experience, it is able to create knowledge. This knowledge is broad, covering any topic that could better an organization. Examples may include ways to increase production efficiency or to develop beneficial investor relations. Knowledge is created at four different units: individual, group, organizational, and inter organizational.

Exam Probability: **Low**

51. *Answer choices:*

(see index for correct answer)

- a. Cooperation
- b. Organization development
- c. Organizational learning
- d. Conflict

Guidance: level 1

:: Stochastic processes ::

_____ in its modern meaning is a "new idea, creative thoughts, new imaginations in form of device or method". _____ is often also viewed as the application of better solutions that meet new requirements, unarticulated needs, or existing market needs. Such _____ takes place through the provision of more-effective products, processes, services, technologies, or business models that are made available to markets, governments and society. An _____ is something original and more effective and, as a consequence, new, that "breaks into" the market or society. _____ is related to, but not the same as, invention, as _____ is more apt to involve the practical implementation of an invention to make a meaningful impact in the market or society, and not all _____ s require an invention. _____ often manifests itself via the engineering process, when the problem being solved is of a technical or scientific nature. The opposite of _____ is exnovation.

Exam Probability: **High**

52. *Answer choices:*

(see index for correct answer)

- a. Narrow escape problem
- b. Hunt process
- c. Bussgang theorem
- d. Bernoulli process

Guidance: level 1

:: ::

A _____ is a research instrument consisting of a series of questions for the purpose of gathering information from respondents. The _____ was invented by the Statistical Society of London in 1838.

Exam Probability: **Low**

53. *Answer choices:*

(see index for correct answer)

- a. Questionnaire
- b. Sarbanes-Oxley act of 2002
- c. personal values
- d. deep-level diversity

Guidance: level 1

:: ::

In business strategy, _____ is establishing a competitive advantage by having the lowest cost of operation in the industry. _____ is often driven by company efficiency, size, scale, scope and cumulative experience .A _____ strategy aims to exploit scale of production, well-defined scope and other economies , producing highly standardized products, using advanced technology.In recent years, more and more companies have chosen a strategic mix to achieve market leadership. These patterns consist of simultaneous _____ , superior customer service and product leadership. Walmart has succeeded across the world due to its _____ strategy. The company has cut down on exesses at every point of production and thus are able to provide the consumers with quality products at low prices.

Exam Probability: **High**

54. *Answer choices:*

(see index for correct answer)

- a. functional perspective
- b. Cost leadership
- c. deep-level diversity
- d. hierarchical perspective

Guidance: level 1

:: Quality management ::

_____ ensures that an organization, product or service is consistent. It has four main components: quality planning, quality assurance, quality control and quality improvement. _____ is focused not only on product and service quality, but also on the means to achieve it. _____ , therefore, uses quality assurance and control of processes as well as products to achieve more consistent quality. What a customer wants and is willing to pay for it determines quality. It is written or unwritten commitment to a known or unknown consumer in the market . Thus, quality can be defined as fitness for intended use or, in other words, how well the product performs its intended function

Exam Probability: **High**

55. *Answer choices:*

(see index for correct answer)

- a. External quality assessment
- b. Quality policy
- c. Quality management
- d. ISO 9000

Guidance: level 1

:: Majority–minority relations ::

_____ , also known as reservation in India and Nepal, positive discrimination / action in the United Kingdom, and employment equity in Canada and South Africa, is the policy of promoting the education and employment of members of groups that are known to have previously suffered from discrimination. Historically and internationally, support for _____ has sought to achieve goals such as bridging inequalities in employment and pay, increasing access to education, promoting diversity, and redressing apparent past wrongs, harms, or hindrances.

Exam Probability: **High**

56. *Answer choices:*
(see index for correct answer)

- a. Affirmative action
- b. positive discrimination
- c. cultural Relativism

Guidance: level 1

:: Mereology ::

_____ , in the abstract, is what belongs to or with something, whether as an attribute or as a component of said thing. In the context of this article, it is one or more components , whether physical or incorporeal, of a person's estate; or so belonging to, as in being owned by, a person or jointly a group of people or a legal entity like a corporation or even a society. Depending on the nature of the _____ , an owner of _____ has the right to consume, alter, share, redefine, rent, mortgage, pawn, sell, exchange, transfer, give away or destroy it, or to exclude others from doing these things, as well as to perhaps abandon it; whereas regardless of the nature of the _____ , the owner thereof has the right to properly use it , or at the very least exclusively keep it.

Exam Probability: **Low**

57. *Answer choices:*

(see index for correct answer)

- a. Simple
- b. Mereology
- c. Gunk
- d. Property

Guidance: level 1

:: Market research ::

_____ is an organized effort to gather information about target markets or customers. It is a very important component of business strategy. The term is commonly interchanged with marketing research; however, expert practitioners may wish to draw a distinction, in that marketing research is concerned specifically about marketing processes, while _____ is concerned specifically with markets.

Exam Probability: **Medium**

58. *Answer choices:*

(see index for correct answer)

- a. Market research
- b. A/B testing
- c. Coolhunting
- d. News ratings in Australia

Guidance: level 1

A _____ is a fund into which a sum of money is added during an employee's employment years, and from which payments are drawn to support the person's retirement from work in the form of periodic payments. A _____ may be a "defined benefit plan" where a fixed sum is paid regularly to a person, or a "defined contribution plan" under which a fixed sum is invested and then becomes available at retirement age. _____ s should not be confused with severance pay; the former is usually paid in regular installments for life after retirement, while the latter is typically paid as a fixed amount after involuntary termination of employment prior to retirement.

Exam Probability: **Medium**

59. *Answer choices:*

(see index for correct answer)

- a. Character
- b. empathy
- c. similarity-attraction theory
- d. Pension

Guidance: level 1

Business law

Corporate law (also known as business law) is the body of law governing the rights, relations, and conduct of persons, companies, organizations and businesses. It refers to the legal practice relating to, or the theory of corporations. Corporate law often describes the law relating to matters which derive directly from the life-cycle of a corporation. It thus encompasses the formation, funding, governance, and death of a corporation.

:: Stock market ::

The _____ of a corporation is all of the shares into which ownership of the corporation is divided. In American English, the shares are commonly known as "_____s". A single share of the _____ represents fractional ownership of the corporation in proportion to the total number of shares. This typically entitles the _____ holder to that fraction of the company's earnings, proceeds from liquidation of assets , or voting power, often dividing these up in proportion to the amount of money each _____ holder has invested. Not all _____ is necessarily equal, as certain classes of _____ may be issued for example without voting rights, with enhanced voting rights, or with a certain priority to receive profits or liquidation proceeds before or after other classes of shareholders.

Exam Probability: **High**

1. *Answer choices:*

(see index for correct answer)

- a. Direct finance
- b. Common stock
- c. Common ordinary equity
- d. Stock

Guidance: level 1

:: Writs ::

In common law, a _____ is a formal _____ ten order issued by a body with administrative or judicial jurisdiction; in modern usage, this body is generally a court. Warrants, prerogative _____ s, and subpoenas are common types of _____ , but many forms exist and have existed.

Exam Probability: **High**

2. *Answer choices:*

(see index for correct answer)

- a. Qui tam
- b. Writ of assistance
- c. Writ of execution

Guidance: level 1

:: ::

A _____ loan or, simply, _____ is used either by purchasers of real property to raise funds to buy real estate, or alternatively by existing property owners to raise funds for any purpose, while putting a lien on the property being _____ d. The loan is "secured" on the borrower's property through a process known as _____ origination. This means that a legal mechanism is put into place which allows the lender to take possession and sell the secured property to pay off the loan in the event the borrower defaults on the loan or otherwise fails to abide by its terms. The word _____ is derived from a Law French term used in Britain in the Middle Ages meaning "death pledge" and refers to the pledge ending when either the obligation is fulfilled or the property is taken through foreclosure. A _____ can also be described as "a borrower giving consideration in the form of a collateral for a benefit ".

Exam Probability: **High**

3. *Answer choices:*

(see index for correct answer)

- a. functional perspective
- b. Sarbanes-Oxley act of 2002
- c. Mortgage
- d. open system

Guidance: level 1

:: ::

In regulatory jurisdictions that provide for it, _____ is a group of laws and organizations designed to ensure the rights of consumers as well as fair trade, competition and accurate information in the marketplace. The laws are designed to prevent the businesses that engage in fraud or specified unfair practices from gaining an advantage over competitors. They may also provides additional protection for those most vulnerable in society. _____ laws are a form of government regulation that aim to protect the rights of consumers. For example, a government may require businesses to disclose detailed information about products—particularly in areas where safety or public health is an issue, such as food.

Exam Probability: **High**

4. *Answer choices:*

(see index for correct answer)

- a. deep-level diversity
- b. hierarchical
- c. Consumer protection
- d. imperative

Guidance: level 1

:: Business law ::

A _____ is a contractual arrangement calling for the lessee to pay the lessor for use of an asset. Property, buildings and vehicles are common assets that are _____ d. Industrial or business equipment is also _____ d.

Exam Probability: **High**

5. *Answer choices:*

(see index for correct answer)

- a. Companies law
- b. Vehicle leasing
- c. Security interest
- d. Business.gov

Guidance: level 1

:: ::

_____ is the consumption and saving opportunity gained by an entity within a specified timeframe, which is generally expressed in monetary terms. For households and individuals, " _____ is the sum of all the wages, salaries, profits, interest payments, rents, and other forms of earnings received in a given period of time."

Exam Probability: **Medium**

6. *Answer choices:*

(see index for correct answer)

- a. Character
- b. hierarchical perspective
- c. Income
- d. corporate values

Guidance: level 1

:: Commercial item transport and distribution ::

_____ s may be negotiable or non-negotiable. Negotiable _____ s allow transfer of ownership of that commodity without having to deliver the physical commodity. See Delivery order.

Exam Probability: **Medium**

7. *Answer choices:*

(see index for correct answer)

- a. Neo-bulk cargo
- b. Bonded warehouse
- c. Warehouse receipt
- d. Currency adjustment factor

Guidance: level 1

:: Real property law ::

> _____, sometimes colloquially described as 'squatter's rights', is a legal principle under which a person who does not have legal title to a piece of property—usually land—acquires legal ownership based on continuous possession or occupation of the land without the permission of its legal owner.

Exam Probability: **Low**

8. *Answer choices:*
(see index for correct answer)

- a. Solar easement
- b. Customary land
- c. Reversionary lease
- d. Adverse possession

Guidance: level 1

:: Business law ::

A _____ is a legal right granted by a debtor to a creditor over the debtor's property which enables the creditor to have recourse to the property if the debtor defaults in making payment or otherwise performing the secured obligations. One of the most common examples of a _____ is a mortgage: When person, by the action of an expressed conveyance, pledges by a promise to pay a certain sum of money, with certain conditions, on a said date or dates for a said period, that action on the page with wet ink applied on the part of the one wishing the exchange creates the original funds and negotiable Instrument. That action of pledging conveys a promise binding upon the mortgagee which creates a face value upon the Instrument of the amount of currency being asked for in exchange. It is therein in good faith offered to the Bank in exchange for local currency from the Bank to buy a house. The particular country's Bank Acts usually requires the Banks to deliver such fund bearing negotiable instruments to the Countries Main Bank such as is the case in Canada. This creates a _____ in the land the house sits on for the Bank and they file a caveat at land titles on the house as evidence of that _____ . If the mortgagee fails to pay defaulting in his promise to repay the exchange, the bank then applies to the court to for-close on your property to eventually sell the house and apply the proceeds to the outstanding exchange.

Exam Probability: **Low**

9. *Answer choices:*

(see index for correct answer)

- a. Security interest
- b. Economic torts
- c. Articles of partnership
- d. Refusal to deal

Guidance: level 1

:: Business law ::

The _____ , first published in 1952, is one of a number of Uniform Acts that have been established as law with the goal of harmonizing the laws of sales and other commercial transactions across the United States of America through UCC adoption by all 50 states, the District of Columbia, and the Territories of the United States.

Exam Probability: **Medium**

10. *Answer choices:*

(see index for correct answer)

- a. De facto corporation and corporation by estoppel
- b. Industrial relations
- c. Stick licensing
- d. Uniform Commercial Code

Guidance: level 1

:: Commercial crimes ::

In law, _____ is the unauthorized use of another's name, likeness, or identity without that person's permission, resulting in harm to that person.

Exam Probability: **High**

11. *Answer choices:*

(see index for correct answer)

- a. Money laundering
- b. The Informant
- c. Gold laundering
- d. Misappropriation

Guidance: level 1

:: ::

Punishment is the imposition of an undesirable or unpleasant outcome upon a group or individual, meted out by an authority—in contexts ranging from child discipline to criminal law—as a response and deterrent to a particular action or behaviour that is deemed undesirable or unacceptable. The reasoning may be to condition a child to avoid self-endangerment, to impose social conformity , to defend norms, to protect against future harms , and to maintain the law—and respect for rule of law—under which the social group is governed. Punishment may be self-inflicted as with self-flagellation and mortification of the flesh in the religious setting, but is most often a form of social coercion.

Exam Probability: **Medium**

12. *Answer choices:*

(see index for correct answer)

- a. Punitive
- b. co-culture

- c. levels of analysis
- d. cultural

Guidance: level 1

:: ::

The _____ of 1977 is a United States federal law known primarily for two of its main provisions: one that addresses accounting transparency requirements under the Securities Exchange Act of 1934 and another concerning bribery of foreign officials. The Act was amended in 1988 and in 1998, and has been subject to continued congressional concerns, namely whether its enforcement discourages U.S. companies from investing abroad.

Exam Probability: **Low**

13. *Answer choices:*

(see index for correct answer)

- a. hierarchical
- b. Foreign Corrupt Practices Act
- c. surface-level diversity
- d. empathy

Guidance: level 1

:: Business law ::

_____ is where a person's financial liability is limited to a fixed sum, most commonly the value of a person's investment in a company or partnership. If a company with _____ is sued, then the claimants are suing the company, not its owners or investors. A shareholder in a limited company is not personally liable for any of the debts of the company, other than for the amount already invested in the company and for any unpaid amount on the shares in the company, if any. The same is true for the members of a _____ partnership and the limited partners in a limited partnership. By contrast, sole proprietors and partners in general partnerships are each liable for all the debts of the business .

Exam Probability: **Low**

14. *Answer choices:*

(see index for correct answer)

- a. Statutory liability
- b. Companies law
- c. Trusted Computing
- d. Limited liability

Guidance: level 1

:: ::

_____ is that part of a civil law legal system which is part of the jus commune that involves relationships between individuals, such as the law of contracts or torts, and the law of obligations. It is to be distinguished from public law, which deals with relationships between both natural and artificial persons and the state, including regulatory statutes, penal law and other law that affects the public order. In general terms, _____ involves interactions between private citizens, whereas public law involves interrelations between the state and the general population.

Exam Probability: **Low**

15. *Answer choices:*

(see index for correct answer)

- a. hierarchical
- b. imperative
- c. Character
- d. Private law

Guidance: level 1

:: Business law ::

A _____, also known as the sole trader, individual entrepreneurship or proprietorship, is a type of enterprise that is owned and run by one person and in which there is no legal distinction between the owner and the business entity. A sole trader does not necessarily work `alone`—it is possible for the sole trader to employ other people.

Exam Probability: **Medium**

16. *Answer choices:*

(see index for correct answer)

- a. Bulk sale
- b. Inslaw
- c. Business license
- d. Sole proprietorship

Guidance: level 1

:: ::

_____ s and acquisitions are transactions in which the ownership of companies, other business organizations, or their operating units are transferred or consolidated with other entities. As an aspect of strategic management, M&A can allow enterprises to grow or downsize, and change the nature of their business or competitive position.

Exam Probability: **High**

17. *Answer choices:*

(see index for correct answer)

- a. functional perspective
- b. information systems assessment

- c. empathy
- d. Merger

Guidance: level 1

:: Negotiable instrument law ::

> In the United States, The Preservation of Consumers' Claims and Defenses [_____ Rule], formally known as the "Trade Regulation Rule Concerning Preservation of Consumers' Claims and Defenses," protects consumers when merchants sell a consumer's credit contracts to other lenders. Specifically, it preserves consumers' right to assert the same legal claims and defenses against anyone who purchases the credit contract, as they would have against the seller who originally provided the credit. [16 Code of Federal Regulations Part 433]

Exam Probability: **High**

18. *Answer choices:*

(see index for correct answer)

- a. Real defense
- b. Swift v. Tyson
- c. Expedited Funds Availability Act
- d. Holder in due course

Guidance: level 1

:: Business models ::

A _____, _____ company or daughter company is a company that is owned or controlled by another company, which is called the parent company, parent, or holding company. The _____ can be a company, corporation, or limited liability company. In some cases it is a government or state-owned enterprise. In some cases, particularly in the music and book publishing industries, subsidiaries are referred to as imprints.

Exam Probability: **Low**

19. *Answer choices:*
(see index for correct answer)

- a. Paid To Click
- b. Subsidiary
- c. Component business model
- d. Lemonade stand

Guidance: level 1

:: Legal doctrines and principles ::

In some common law jurisdictions, _____ is a defense to a tort claim based on negligence. If it is available, the defense completely bars plaintiffs from any recovery if they contribute to their own injury through their own negligence.

Exam Probability: **Medium**

20. *Answer choices:*

(see index for correct answer)

- a. Act of state
- b. unconscionable contract
- c. Contributory negligence
- d. Attractive nuisance doctrine

Guidance: level 1

:: ::

The _____ is the highest court within the hierarchy of courts in many legal jurisdictions. Other descriptions for such courts include court of last resort, apex court, and high court of appeal. Broadly speaking, the decisions of a _____ are not subject to further review by any other court. _____ s typically function primarily as appellate courts, hearing appeals from decisions of lower trial courts, or from intermediate-level appellate courts.

Exam Probability: **High**

21. *Answer choices:*

(see index for correct answer)

- a. hierarchical perspective

- b. Supreme Court
- c. levels of analysis
- d. cultural

Guidance: level 1

:: Contract law ::

Generally, a _____ is a loan or a credit transaction in which the lender acquires a security interest in collateral owned by the borrower and is entitled to foreclose on or repossess the collateral in the event of the borrower's default. The terms of the relationship are governed by a contract, or security agreement. A common example would be a consumer who purchases a car on credit. If the consumer fails to make the payments on time, the lender will take the car and resell it, applying the proceeds of the sale toward the loan. Mortgages and deeds of trust are another example. In the United States, _____ s in personal property are governed by Article 9 of the Uniform Commercial Code .

Exam Probability: **Low**

22. *Answer choices:*

(see index for correct answer)

- a. Secured transaction
- b. Estoppel by deed
- c. Offeree
- d. Force majeure

Guidance: level 1

:: ::

_____ or accountancy is the measurement, processing, and communication of financial information about economic entities such as businesses and corporations. The modern field was established by the Italian mathematician Luca Pacioli in 1494. _____ , which has been called the "language of business", measures the results of an organization's economic activities and conveys this information to a variety of users, including investors, creditors, management, and regulators. Practitioners of _____ are known as accountants. The terms " _____ " and "financial reporting" are often used as synonyms.

Exam Probability: **High**

23. *Answer choices:*

(see index for correct answer)

- a. Accounting
- b. similarity-attraction theory
- c. open system
- d. co-culture

Guidance: level 1

:: Business ::

_____ is a trade policy that does not restrict imports or exports; it can also be understood as the free market idea applied to international trade. In government, _____ is predominantly advocated by political parties that hold liberal economic positions while economically left-wing and nationalist political parties generally support protectionism, the opposite of _____ .

Exam Probability: **High**

24. *Answer choices:*

(see index for correct answer)

- a. Sales
- b. Free trade
- c. Kingdomality
- d. Employee experience management

Guidance: level 1

:: Criminal procedure ::

In law, a verdict is the formal finding of fact made by a jury on matters or questions submitted to the jury by a judge. In a bench trial, the judge's decision near the end of the trial is simply referred to as a finding. In England and Wales, a coroner's findings are called verdicts .

Exam Probability: **High**

25. *Answer choices:*

(see index for correct answer)

- a. criminal procedure
- b. Exoneration

Guidance: level 1

:: Insurance terms ::

> A _____ in the broadest sense is a natural person or other legal entity who receives money or other benefits from a benefactor. For example, the _____ of a life insurance policy is the person who receives the payment of the amount of insurance after the death of the insured.

Exam Probability: **High**

26. *Answer choices:*

(see index for correct answer)

- a. Beneficiary
- b. Short Rate Table
- c. replacement cost
- d. Probable maximum loss

Guidance: level 1

:: ::

An _____ is a contingent motivator. Traditional _____ s are extrinsic motivators which reward actions to yield a desired outcome. The effectiveness of traditional _____ s has changed as the needs of Western society have evolved. While the traditional _____ model is effective when there is a defined procedure and goal for a task, Western society started to require a higher volume of critical thinkers, so the traditional model became less effective. Institutions are now following a trend in implementing strategies that rely on intrinsic motivations rather than the extrinsic motivations that the traditional _____ s foster.

Exam Probability: **Medium**

27. *Answer choices:*

(see index for correct answer)

- a. Incentive
- b. hierarchical perspective
- c. imperative
- d. information systems assessment

Guidance: level 1

:: Abuse of the legal system ::

_____ occurs when a person is restricted in their personal movement within any area without justification or consent. Actual physical restraint is not necessary for _____ to occur. A _____ claim may be made based upon private acts, or upon wrongful governmental detention. For detention by the police, proof of _____ provides a basis to obtain a writ of habeas corpus.

Exam Probability: **High**

28. *Answer choices:*

(see index for correct answer)

- a. Obstruction of Justice
- b. False imprisonment
- c. Forum shopping

Guidance: level 1

:: ::

_____ , or auditory perception, is the ability to perceive sounds by detecting vibrations, changes in the pressure of the surrounding medium through time, through an organ such as the ear. The academic field concerned with _____ is auditory science.

Exam Probability: **Low**

29. Answer choices:

(see index for correct answer)

- a. hierarchical perspective
- b. Hearing
- c. corporate values
- d. cultural

Guidance: level 1

:: Contract law ::

A _____ cannot be enforced by law. _____ s are different from voidable contracts, which are contracts that may be nullified. However, when a contract is being written and signed, there is no automatic mechanism available in every situation that can be utilized to detect the validity or enforceability of that contract. Practically, a contract can be declared to be void by a court of law. So the main question is that under what conditions can a contract be deemed as void

Exam Probability: **High**

30. Answer choices:

(see index for correct answer)

- a. Void contract
- b. Convention on the Law Applicable to Contractual Obligations 1980
- c. Scots contract law

- d. Unenforceable

Guidance: level 1

:: Contract law ::

Coercion is the practice of forcing another party to act in an involuntary manner by use of threats or force. It involves a set of various types of forceful actions that violate the free will of an individual to induce a desired response, for example: a bully demanding lunch money from a student or the student gets beaten. These actions may include extortion, blackmail, torture, threats to induce favors, or even sexual assault. In law, coercion is codified as a _____ crime. Such actions are used as leverage, to force the victim to act in a way contrary to their own interests. Coercion may involve the actual infliction of physical pain/injury or psychological harm in order to enhance the credibility of a threat. The threat of further harm may lead to the cooperation or obedience of the person being coerced.

Exam Probability: **Medium**

31. *Answer choices:*
(see index for correct answer)

- a. Neo-classical contract
- b. Prenuptial agreement
- c. Bonus clause
- d. Duress

Guidance: level 1

:: Contract law ::

Offer and acceptance analysis is a traditional approach in contract law. The offer and acceptance formula, developed in the 19th century, identifies a moment of formation when the parties are of one mind. This classical approach to contract formation has been modified by developments in the law of estoppel, misleading conduct, misrepresentation and unjust enrichment.

Exam Probability: **Low**

32. *Answer choices:*
(see index for correct answer)

- a. Interlineation
- b. Unenforceable
- c. Job order contracting
- d. Indian contract law

Guidance: level 1

:: Jurisdiction ::

In United States law, _____ jurisdiction is the subject-matter jurisdiction of United States federal courts to hear a civil case because the plaintiff has alleged a violation of the United States Constitution, federal law, or a treaty to which the United States is a party.

Exam Probability: **Low**

33. *Answer choices:*

(see index for correct answer)

- a. concurrent jurisdiction
- b. Federal question
- c. General jurisdiction
- d. Jurisdiction in rem

Guidance: level 1

:: Latin legal terms ::

In law and government, _____ describes practices that are legally recognised, regardless whether the practice exists in reality. In contrast, de facto describes situations that exist in reality, even if not legally recognised. The terms are often used to contrast different scenarios: for a colloquial example, "I know that, _____ , this is supposed to be a parking lot, but now that the flood has left four feet of water here, it's a de facto swimming pool". To further explain, even if the signs around the flooded parking lot say "Parking Lot" it is "in fact" a swimming pool.

Exam Probability: **Low**

34. *Answer choices:*

(see index for correct answer)

- a. Quid pro quo
- b. de facto
- c. De jure
- d. Nolo contendere

Guidance: level 1

:: Finance ::

A _____ , in the law of the United States, is a contract that governs the relationship between the parties to a kind of financial transaction known as a secured transaction. In a secured transaction, the Grantor assigns, grants and pledges to the grantee a security interest in personal property which is referred to as the collateral. Examples of typical collateral are shares of stock, livestock, and vehicles. A _____ is not used to transfer any interest in real property , only personal property. The document used by lenders to obtain a lien on real property is a mortgage or deed of trust.

Exam Probability: **High**

35. *Answer choices:*
(see index for correct answer)

- a. Liabilities Subject to Compromise
- b. Operating partner
- c. Target benefit plan
- d. Security agreement

Guidance: level 1

:: Chemical industry ::

The _____ for the Protection of Literary and Artistic Works, usually known as the _____, is an international agreement governing copyright, which was first accepted in Berne, Switzerland, in 1886.

Exam Probability: **Low**

36. *Answer choices:*

(see index for correct answer)

- a. Carl Kellner
- b. IGCW
- c. Berne Convention
- d. Chemical plant

Guidance: level 1

:: Real estate ::

_____ , real estate, realty, or immovable property In English common law refers to landed properties belonging to some person. It include all structures, crops, buildings, machinery, wells, dams, ponds, mines, canals, and roads, among other things. The term is historic, arising from the now-discontinued form of action, which distinguish between _____ disputes and personal property disputes. Personal property was, and continues to refer to all properties that are not real properties.

Exam Probability: **High**

37. *Answer choices:*

<small>(see index for correct answer)</small>

- a. Privately owned public space
- b. Chambre de bonne
- c. Originating application
- d. Jay Papasan

Guidance: level 1

:: Contract law ::

_____ is a legal process for collecting a monetary judgment on behalf of a plaintiff from a defendant. _____ allows the plaintiff to take the money or property of the debtor from the person or institution that holds that property. A similar legal mechanism called execution allows the seizure of money or property held directly by the debtor.

Exam Probability: **High**

38. *Answer choices:*

(see index for correct answer)

- a. Pact ink
- b. Garnishment
- c. Posting rule
- d. Four corners

Guidance: level 1

:: ::

> A concept of English law, a _____ is an untrue or misleading statement of fact made during negotiations by one party to another, the statement then inducing that other party into the contract. The misled party may normally rescind the contract, and sometimes may be awarded damages as well.

Exam Probability: **Low**

39. *Answer choices:*

(see index for correct answer)

- a. surface-level diversity
- b. Misrepresentation

- c. Character
- d. similarity-attraction theory

Guidance: level 1

:: Criminal law ::

_____ is the body of law that relates to crime. It proscribes conduct perceived as threatening, harmful, or otherwise endangering to the property, health, safety, and moral welfare of people inclusive of one's self. Most _____ is established by statute, which is to say that the laws are enacted by a legislature. _____ includes the punishment and rehabilitation of people who violate such laws. _____ varies according to jurisdiction, and differs from civil law, where emphasis is more on dispute resolution and victim compensation, rather than on punishment or rehabilitation. Criminal procedure is a formalized official activity that authenticates the fact of commission of a crime and authorizes punitive or rehabilitative treatment of the offender.

Exam Probability: **Low**

40. *Answer choices:*

(see index for correct answer)

- a. Mala prohibita
- b. complicit
- c. Mala in se
- d. mitigating factor

Guidance: level 1

:: Contract law ::

_____ is a doctrine in contract law that describes terms that are so extremely unjust, or overwhelmingly one-sided in favor of the party who has the superior bargaining power, that they are contrary to good conscience. Typically, an unconscionable contract is held to be unenforceable because no reasonable or informed person would otherwise agree to it. The perpetrator of the conduct is not allowed to benefit, because the consideration offered is lacking, or is so obviously inadequate, that to enforce the contract would be unfair to the party seeking to escape the contract.

Exam Probability: **High**

41. *Answer choices:*

(see index for correct answer)

- a. Unconscionability
- b. Efficient breach
- c. The Death of Contract
- d. Warranty tolling

Guidance: level 1

:: Mereology ::

_____, in the abstract, is what belongs to or with something, whether as an attribute or as a component of said thing. In the context of this article, it is one or more components, whether physical or incorporeal, of a person's estate; or so belonging to, as in being owned by, a person or jointly a group of people or a legal entity like a corporation or even a society. Depending on the nature of the _____, an owner of _____ has the right to consume, alter, share, redefine, rent, mortgage, pawn, sell, exchange, transfer, give away or destroy it, or to exclude others from doing these things, as well as to perhaps abandon it; whereas regardless of the nature of the _____, the owner thereof has the right to properly use it, or at the very least exclusively keep it.

Exam Probability: **Medium**

42. *Answer choices:*

(see index for correct answer)

- a. Mereological nihilism
- b. Mereological essentialism
- c. Property
- d. Mereotopology

Guidance: level 1

:: ::

_____ is a judicial device in common law legal systems whereby a court may prevent, or "estop" a person from making assertions or from going back on his or her word; the person being sanctioned is "estopped". _____ may prevent someone from bringing a particular claim. Legal doctrines of _____ are based in both common law and equity.

Exam Probability: **Low**

43. *Answer choices:*
(see index for correct answer)

- a. similarity-attraction theory
- b. hierarchical perspective
- c. Estoppel
- d. Sarbanes-Oxley act of 2002

Guidance: level 1

:: Contract law ::

A _____ is a legally-binding agreement which recognises and governs the rights and duties of the parties to the agreement. A _____ is legally enforceable because it meets the requirements and approval of the law. An agreement typically involves the exchange of goods, services, money, or promises of any of those. In the event of breach of _____ , the law awards the injured party access to legal remedies such as damages and cancellation.

Exam Probability: **Low**

44. *Answer choices:*

(see index for correct answer)

- a. Baseball business rules
- b. Seaworthiness
- c. Contract
- d. Perfect tender

Guidance: level 1

:: Marketing ::

_____ or stock is the goods and materials that a business holds for the ultimate goal of resale.

Exam Probability: **High**

45. *Answer choices:*

(see index for correct answer)

- a. Instant rebate
- b. John Neeson
- c. Exploratory research
- d. Inventory

Guidance: level 1

:: Financial regulatory authorities of the United States ::

The _____ is the revenue service of the United States federal government. The government agency is a bureau of the Department of the Treasury, and is under the immediate direction of the Commissioner of Internal Revenue, who is appointed to a five-year term by the President of the United States. The IRS is responsible for collecting taxes and administering the Internal Revenue Code, the main body of federal statutory tax law of the United States. The duties of the IRS include providing tax assistance to taxpayers and pursuing and resolving instances of erroneous or fraudulent tax filings. The IRS has also overseen various benefits programs, and enforces portions of the Affordable Care Act.

Exam Probability: **Low**

46. *Answer choices:*
(see index for correct answer)

- a. U.S. Securities and Exchange Commission
- b. Municipal Securities Rulemaking Board
- c. National Credit Union Administration
- d. Internal Revenue Service

Guidance: level 1

:: ::

_____ is the practical authority granted to a legal body to administer justice within a defined field of responsibility, e.g., Michigan tax law. In federations like the United States, areas of _____ apply to local, state, and federal levels; e.g. the court has _____ to apply federal law.

Exam Probability: **Medium**

47. *Answer choices:*

(see index for correct answer)

- a. similarity-attraction theory
- b. functional perspective
- c. cultural
- d. process perspective

Guidance: level 1

:: Stock market ::

_____ is freedom from, or resilience against, potential harm caused by others. Beneficiaries of _____ may be of persons and social groups, objects and institutions, ecosystems or any other entity or phenomenon vulnerable to unwanted change by its environment.

Exam Probability: **Medium**

48. *Answer choices:*

(see index for correct answer)

- a. Security
- b. Burgundy
- c. Control premium
- d. Contract for difference

Guidance: level 1

:: Contract law ::

_____ is an equitable remedy in the law of contract, whereby a court issues an order requiring a party to perform a specific act, such as to complete performance of the contract. It is typically available in the sale of land, but otherwise is not generally available if damages are an appropriate alternative. _____ is almost never available for contracts of personal service, although performance may also be ensured through the threat of proceedings for contempt of court.

Exam Probability: **Medium**

49. *Answer choices:*

(see index for correct answer)

- a. Secured transaction
- b. Contingent contracts
- c. Impracticability
- d. Piggy-back

Guidance: level 1

:: Competition law ::

In competition law, a _____ is a market in which a particular product or service is sold. It is the intersection of a relevant product market and a relevant geographic market. The European Commission defines a _____ and its product and geographic components as follows.

Exam Probability: **Low**

50. *Answer choices:*

(see index for correct answer)

- a. Relevant market
- b. European Union merger law
- c. Orange-Book-Standard
- d. International Competition Network

Guidance: level 1

:: Commercial crimes ::

_____ is the process of concealing the origins of money obtained illegally by passing it through a complex sequence of banking transfers or commercial transactions. The overall scheme of this process returns the money to the launderer in an obscure and indirect way.

Exam Probability: **Medium**

51. *Answer choices:*

(see index for correct answer)

- a. Late trading
- b. Counterfeit consumer goods
- c. Terrorism financing
- d. Gold laundering

Guidance: level 1

:: Business law ::

The term is used to designate a range of diverse, if often kindred, concepts. These have historically been addressed in a number of discrete disciplines, notably mathematics, physics, chemistry, ethics, aesthetics, ontology, and theology.

Exam Probability: **Low**

52. *Answer choices:*

(see index for correct answer)

- a. Contract A
- b. Lessor
- c. Court auction
- d. Perfection

Guidance: level 1

:: Real property law ::

A _____ is any legal instrument in writing which passes, affirms or confirms an interest, right, or property and that is signed, attested, delivered, and in some jurisdictions, sealed. It is commonly associated with transferring title to property. The _____ has a greater presumption of validity and is less rebuttable than an instrument signed by the party to the _____ . A _____ can be unilateral or bilateral. _____ s include conveyances, commissions, licenses, patents, diplomas, and conditionally powers of attorney if executed as _____ s. The _____ is the modern descendant of the medieval charter, and delivery is thought to symbolically replace the ancient ceremony of livery of seisin.

Exam Probability: **High**

53. *Answer choices:*
(see index for correct answer)

- a. Deed
- b. Fee farm grant

- c. Doctrine of worthier title
- d. Chalking the door

Guidance: level 1

:: Mortgage ::

_____ is a legal process in which a lender attempts to recover the balance of a loan from a borrower who has stopped making payments to the lender by forcing the sale of the asset used as the collateral for the loan.

Exam Probability: **Low**

54. *Answer choices:*

(see index for correct answer)

- a. Foreclosure
- b. Flexible mortgage
- c. Fractional financing
- d. Second mortgage

Guidance: level 1

:: Fair use ::

_____ is a doctrine in the law of the United States that permits limited use of copyrighted material without having to first acquire permission from the copyright holder. _____ is one of the limitations to copyright intended to balance the interests of copyright holders with the public interest in the wider distribution and use of creative works by allowing as a defense to copyright infringement claims certain limited uses that might otherwise be considered infringement.

Exam Probability: **High**

55. *Answer choices:*

(see index for correct answer)

- a. FAIR USE Act
- b. Fair use
- c. Transformation
- d. Toward a Fair Use Standard

Guidance: level 1

:: False advertising law ::

The Lanham Act is the primary federal trademark statute of law in the United States. The Act prohibits a number of activities, including trademark infringement, trademark dilution, and false advertising.

Exam Probability: **High**

56. Answer choices:

(see index for correct answer)

- a. Rebecca Tushnet
- b. POM Wonderful LLC v. Coca-Cola Co.

Guidance: level 1

:: Contract law ::

_____ is a legal cause of action and a type of civil wrong, in which a binding agreement or bargained-for exchange is not honored by one or more of the parties to the contract by non-performance or interference with the other party's performance. Breach occurs when a party to a contract fails to fulfill its obligation as described in the contract, or communicates an intent to fail the obligation or otherwise appears not to be able to perform its obligation under the contract. Where there is _____, the resulting damages will have to be paid by the party breaching the contract to the aggrieved party.

Exam Probability: **High**

57. Answer choices:

(see index for correct answer)

- a. Estoppel by deed
- b. Handshake deal
- c. Warranty tolling
- d. Meeting of the minds

Guidance: level 1

:: ::

The _____ is an independent agency of the Federal government of the United States with responsibilities for enforcing U.S. labor law in relation to collective bargaining and unfair labor practices. Under the National Labor Relations Act of 1935 it supervises elections for labor union representation and can investigate and remedy unfair labor practices. Unfair labor practices may involve union-related situations or instances of protected concerted activity. The NLRB is governed by a five-person board and a General Counsel, all of whom are appointed by the President with the consent of the Senate. Board members are appointed to five-year terms and the General Counsel is appointed to a four-year term. The General Counsel acts as a prosecutor and the Board acts as an appellate quasi-judicial body from decisions of administrative law judges.

Exam Probability: **High**

58. *Answer choices:*

(see index for correct answer)

- a. functional perspective
- b. cultural
- c. process perspective
- d. personal values

Guidance: level 1

:: Legal doctrines and principles ::

_____ is a doctrine that a party is responsible for acts of their agents. For example, in the United States, there are circumstances when an employer is liable for acts of employees performed within the course of their employment. This rule is also called the master-servant rule, recognized in both common law and civil law jurisdictions.

Exam Probability: **Medium**

59. *Answer choices:*

(see index for correct answer)

- a. Acquiescence
- b. Mutual assent
- c. Duty to rescue
- d. Respondeat superior

Guidance: level 1

Finance

Finance is a field that is concerned with the allocation (investment) of assets and liabilities over space and time, often under conditions of risk or uncertainty. Finance can also be defined as the science of money management. Participants in the market aim to price assets based on their risk level, fundamental value, and their expected rate of return. Finance can be split into three sub-categories: public finance, corporate finance and personal finance.

:: Costs ::

_____ is the sum of costs of all resources consumed in the process of making a product. The _____ is classified into three categories: direct materials cost, direct labor cost and manufacturing overhead.

Exam Probability: **Medium**

1. *Answer choices:*

(see index for correct answer)

- a. Manufacturing cost
- b. Opportunity cost of capital
- c. Flyaway cost
- d. Average variable cost

Guidance: level 1

:: Financial ratios ::

A _____ or accounting ratio is a relative magnitude of two selected numerical values taken from an enterprise's financial statements. Often used in accounting, there are many standard ratios used to try to evaluate the overall financial condition of a corporation or other organization. _____ s may be used by managers within a firm, by current and potential shareholders of a firm, and by a firm's creditors. Financial analysts use _____ s to compare the strengths and weaknesses in various companies. If shares in a company are traded in a financial market, the market price of the shares is used in certain _____ s.

Exam Probability: **Low**

2. *Answer choices:*

(see index for correct answer)

- a. PE ratio
- b. Theoretical ex-rights price
- c. Financial ratio
- d. Return on event

Guidance: level 1

:: Asset ::

In financial accounting, an _____ is any resource owned by the business. Anything tangible or intangible that can be owned or controlled to produce value and that is held by a company to produce positive economic value is an _____. Simply stated, _____ s represent value of ownership that can be converted into cash. The balance sheet of a firm records the monetary value of the _____ s owned by that firm. It covers money and other valuables belonging to an individual or to a business.

Exam Probability: **Low**

3. *Answer choices:*

(see index for correct answer)

- a. Fixed asset
- b. Asset

Guidance: level 1

:: Margin policy ::

> In finance, a _____ is a standardized forward contract, a legal agreement to buy or sell something at a predetermined price at a specified time in the future, between parties not known to each other. The asset transacted is usually a commodity or financial instrument. The predetermined price the parties agree to buy and sell the asset for is known as the forward price. The specified time in the future—which is when delivery and payment occur—is known as the delivery date. Because it is a function of an underlying asset, a _____ is a derivative product.

Exam Probability: **High**

4. *Answer choices:*

(see index for correct answer)

- a. Futures contract
- b. Regulation T

Guidance: level 1

:: ::

An _____ is a comprehensive report on a company's activities throughout the preceding year. _____ s are intended to give shareholders and other interested people information about the company's activities and financial performance. They may be considered as grey literature. Most jurisdictions require companies to prepare and disclose _____ s, and many require the _____ to be filed at the company's registry. Companies listed on a stock exchange are also required to report at more frequent intervals .

Exam Probability: **High**

5. *Answer choices:*

(see index for correct answer)

- a. Character
- b. Sarbanes-Oxley act of 2002
- c. imperative
- d. Annual report

Guidance: level 1

:: Accounting terminology ::

_____ is a legally enforceable claim for payment held by a business for goods supplied and/or services rendered that customers/clients have ordered but not paid for. These are generally in the form of invoices raised by a business and delivered to the customer for payment within an agreed time frame. _____ is shown in a balance sheet as an asset. It is one of a series of accounting transactions dealing with the billing of a customer for goods and services that the customer has ordered. These may be distinguished from notes receivable, which are debts created through formal legal instruments called promissory notes.

Exam Probability: **High**

6. *Answer choices:*

(see index for correct answer)

- a. Total absorption costing
- b. Accounts receivable
- c. Accrued liabilities
- d. Accounting equation

Guidance: level 1

:: Mutualism (movement) ::

A _____ is a professionally managed investment fund that pools money from many investors to purchase securities. These investors may be retail or institutional in nature.

Exam Probability: **High**

7. *Answer choices:*

(see index for correct answer)

- a. Mutualism
- b. Sovereigns of Industry
- c. Benefit society
- d. Mutual fund

Guidance: level 1

:: Land value taxation ::

_____ , sometimes referred to as dry _____ , is the solid surface of Earth that is not permanently covered by water. The vast majority of human activity throughout history has occurred in _____ areas that support agriculture, habitat, and various natural resources. Some life forms have developed from predecessor species that lived in bodies of water.

Exam Probability: **Low**

8. *Answer choices:*

(see index for correct answer)

- a. Georgism
- b. Harry Gunnison Brown

- c. Land value tax
- d. Lands Valuation Appeal Court

Guidance: level 1

:: Generally Accepted Accounting Principles ::

In business and accounting, _____ is an entity's income minus cost of goods sold, expenses and taxes for an accounting period. It is computed as the residual of all revenues and gains over all expenses and losses for the period, and has also been defined as the net increase in shareholders' equity that results from a company's operations. In the context of the presentation of financial statements, the IFRS Foundation defines _____ as synonymous with profit and loss. The difference between revenue and the cost of making a product or providing a service, before deducting overheads, payroll, taxation, and interest payments. This is different from operating income.

Exam Probability: **Low**

9. *Answer choices:*
(see index for correct answer)

- a. Chinese accounting standards
- b. Net income
- c. net realisable value
- d. Write-off

Guidance: level 1

:: Market research ::

_____ , an acronym for Information through Disguised Experimentation is an annual market research fair conducted by the students of IIM-Lucknow. Students create games and use various other simulated environments to capture consumers' subconscious thoughts. This innovative method of market research removes the sensitization effect that might bias peoples answers to questions. This ensures that the most truthful answers are captured to research questions. The games are designed in such a way that the observers can elicit all the required information just by observing and noting down the behaviour and the responses of the participants.

Exam Probability: **Medium**

10. *Answer choices:*

(see index for correct answer)

- a. Friday night death slot
- b. Automated Measurement of Lineups
- c. INDEX
- d. Market surveillance

Guidance: level 1

:: Fixed income market ::

In finance, the _____ is a curve showing several yields or interest rates across different contract lengths for a similar debt contract. The curve shows the relation between the interest rate and the time to maturity, known as the "term", of the debt for a given borrower in a given currency. For example, the U.S. dollar interest rates paid on U.S. Treasury securities for various maturities are closely watched by many traders, and are commonly plotted on a graph such as the one on the right which is informally called "the _____ ". More formal mathematical descriptions of this relation are often called the term structure of interest rates.

Exam Probability: **Medium**

11. *Answer choices:*
(see index for correct answer)

- a. Basis point
- b. Yield curve
- c. Fixed-income attribution
- d. Fixed income

Guidance: level 1

:: Human resource management ::

_____ is the corporate management term for the act of reorganizing the legal, ownership, operational, or other structures of a company for the purpose of making it more profitable, or better organized for its present needs. Other reasons for _____ include a change of ownership or ownership structure, demerger, or a response to a crisis or major change in the business such as bankruptcy, repositioning, or buyout. _____ may also be described as corporate _____, debt _____ and financial _____.

Exam Probability: **Medium**

12. *Answer choices:*

(see index for correct answer)

- a. Individual development plan
- b. Hemsley Fraser
- c. war for talent
- d. Professional employer organization

Guidance: level 1

:: Stock market ::

_____ or stock market launch is a type of public offering in which shares of a company are sold to institutional investors and usually also retail investors; an IPO is underwritten by one or more investment banks, who also arrange for the shares to be listed on one or more stock exchanges. Through this process, colloquially known as floating, or going public, a privately held company is transformed into a public company. _____ s can be used: to raise new equity capital for the company concerned; to monetize the investments of private shareholders such as company founders or private equity investors; and to enable easy trading of existing holdings or future capital raising by becoming publicly traded enterprises.

Exam Probability: **High**

13. *Answer choices:*

(see index for correct answer)

- a. Red chip
- b. Paper valuation
- c. Seeking Alpha
- d. Initial public offering

Guidance: level 1

:: Generally Accepted Accounting Principles ::

A _____ , in accrual accounting, is any account where the asset or liability is not realized until a future date , e.g. annuities, charges, taxes, income, etc. The deferred item may be carried, dependent on type of _____ , as either an asset or liability. See also accrual.

Exam Probability: **High**

14. *Answer choices:*

(see index for correct answer)

- a. Deferred income
- b. Generally Accepted Accounting Practice
- c. Profit
- d. Deferral

Guidance: level 1

:: ::

_____ is the production of products for use or sale using labour and machines, tools, chemical and biological processing, or formulation. The term may refer to a range of human activity, from handicraft to high tech, but is most commonly applied to industrial design, in which raw materials are transformed into finished goods on a large scale. Such finished goods may be sold to other manufacturers for the production of other, more complex products, such as aircraft, household appliances, furniture, sports equipment or automobiles, or sold to wholesalers, who in turn sell them to retailers, who then sell them to end users and consumers.

Exam Probability: **Medium**

15. *Answer choices:*

(see index for correct answer)

- a. levels of analysis
- b. Manufacturing
- c. hierarchical perspective
- d. information systems assessment

Guidance: level 1

:: Options (finance) ::

A _____ , often simply labeled a "call", is a financial contract between two parties, the buyer and the seller of this type of option. The buyer of the _____ has the right, but not the obligation, to buy an agreed quantity of a particular commodity or financial instrument from the seller of the option at a certain time for a certain price . The seller is obligated to sell the commodity or financial instrument to the buyer if the buyer so decides. The buyer pays a fee for this right. The term "call" comes from the fact that the owner has the right to "call the stock away" from the seller.

Exam Probability: **Medium**

16. *Answer choices:*

(see index for correct answer)

- a. Timer Call
- b. callable
- c. Turbo warrant
- d. Call option

Guidance: level 1

:: Management accounting ::

_____ accounting is a traditional cost accounting method introduced in the 1920s, as an alternative for the traditional cost accounting method based on historical costs.

Exam Probability: **Low**

17. *Answer choices:*

(see index for correct answer)

- a. Spend management
- b. Job costing
- c. Revenue center
- d. Standard cost

Guidance: level 1

:: ::

The _____ is a private, non-profit organization standard-setting body whose primary purpose is to establish and improve Generally Accepted Accounting Principles within the United States in the public's interest. The Securities and Exchange Commission designated the FASB as the organization responsible for setting accounting standards for public companies in the US. The FASB replaced the American Institute of Certified Public Accountants` Accounting Principles Board on July 1, 1973.

Exam Probability: **Low**

18. *Answer choices:*

(see index for correct answer)

- a. cultural
- b. Character
- c. empathy
- d. Financial Accounting Standards Board

Guidance: level 1

:: Materials ::

A _____ , also known as a feedstock, unprocessed material, or primary commodity, is a basic material that is used to produce goods, finished products, energy, or intermediate materials which are feedstock for future finished products. As feedstock, the term connotes these materials are bottleneck assets and are highly important with regard to producing other products. An example of this is crude oil, which is a _____ and a feedstock used in the production of industrial chemicals, fuels, plastics, and pharmaceutical goods; lumber is a _____ used to produce a variety of products including all types of furniture. The term "_____" denotes materials in minimally processed or unprocessed in states; e.g., raw latex, crude oil, cotton, coal, raw biomass, iron ore, air, logs, or water i.e. "...any product of agriculture, forestry, fishing and any other mineral that is in its natural form or which has undergone the transformation required to prepare it for internationally marketing in substantial volumes."

Exam Probability: **High**

19. *Answer choices:*

(see index for correct answer)

- a. Tortoiseshell
- b. Semimetal
- c. Raw material
- d. Rubblization

Guidance: level 1

:: Currency ::

A _____ , in the most specific sense is money in any form when in use or circulation as a medium of exchange, especially circulating banknotes and coins. A more general definition is that a _____ is a system of money in common use, especially for people in a nation. Under this definition, US dollars , pounds sterling , Australian dollars , European euros , Russian rubles and Indian Rupees are examples of currencies. These various currencies are recognized as stores of value and are traded between nations in foreign exchange markets, which determine the relative values of the different currencies. Currencies in this sense are defined by governments, and each type has limited boundaries of acceptance.

Exam Probability: **Low**

20. *Answer choices:*

(see index for correct answer)

- a. Currency money
- b. Circulation
- c. Currency
- d. Currency basket

Guidance: level 1

:: Generally Accepted Accounting Principles ::

_____ is the accounting classification of an account. It is part of double-entry book-keeping technique.

Exam Probability: **Medium**

21. *Answer choices:*

(see index for correct answer)

- a. deferred revenue
- b. Revenue
- c. Profit
- d. Normal balance

Guidance: level 1

:: Business law ::

A _____ is an arrangement where parties, known as partners, agree to cooperate to advance their mutual interests. The partners in a _____ may be individuals, businesses, interest-based organizations, schools, governments or combinations. Organizations may partner to increase the likelihood of each achieving their mission and to amplify their reach. A _____ may result in issuing and holding equity or may be only governed by a contract.

Exam Probability: **Low**

22. *Answer choices:*

(see index for correct answer)

- a. Partnership
- b. Unfair Commercial Practices Directive

- c. Process agent
- d. Lessor

Guidance: level 1

:: Financial ratios ::

> The _____ or dividend-price ratio of a share is the dividend per share, divided by the price per share. It is also a company's total annual dividend payments divided by its market capitalization, assuming the number of shares is constant. It is often expressed as a percentage.

Exam Probability: **High**

23. *Answer choices:*
(see index for correct answer)

- a. Diluted earnings per share
- b. Dividend yield
- c. Bias ratio
- d. PE ratio

Guidance: level 1

:: United States Generally Accepted Accounting Principles ::

In a companies' financial reporting, _____ "includes all changes in equity during a period except those resulting from investments by owners and distributions to owners". Because that use excludes the effects of changing ownership interest, an economic measure of _____ is necessary for financial analysis from the shareholders' point of view

Exam Probability: **Low**

24. *Answer choices:*

(see index for correct answer)

- a. Available for sale
- b. Single Audit
- c. Comprehensive income
- d. GASB 45

Guidance: level 1

:: ::

_____ refers to a business or organization attempting to acquire goods or services to accomplish its goals. Although there are several organizations that attempt to set standards in the _____ process, processes can vary greatly between organizations. Typically the word " _____ " is not used interchangeably with the word "procurement", since procurement typically includes expediting, supplier quality, and transportation and logistics in addition to _____ .

Exam Probability: **Medium**

25. *Answer choices:*

(see index for correct answer)

- a. empathy
- b. imperative
- c. process perspective
- d. hierarchical perspective

Guidance: level 1

:: ::

A _____ is the period used by governments for accounting and budget purposes, which varies between countries. It is also used for financial reporting by business and other organizations. Laws in many jurisdictions require company financial reports to be prepared and published on an annual basis, but generally do not require the reporting period to align with the calendar year . Taxation laws generally require accounting records to be maintained and taxes calculated on an annual basis, which usually corresponds to the _____ used for government purposes. The calculation of tax on an annual basis is especially relevant for direct taxation, such as income tax. Many annual government fees—such as Council rates, licence fees, etc.—are also levied on a _____ basis, while others are charged on an anniversary basis.

Exam Probability: **Medium**

26. *Answer choices:*

(see index for correct answer)

- a. co-culture
- b. information systems assessment
- c. Character
- d. personal values

Guidance: level 1

:: Consumer theory ::

A _____ is a technical term in psychology, economics and philosophy usually used in relation to choosing between alternatives. For example, someone prefers A over B if they would rather choose A than B.

Exam Probability: **High**

27. *Answer choices:*

(see index for correct answer)

- a. Business contract hire
- b. Marshallian demand function
- c. Permanent income hypothesis
- d. Preference

Guidance: level 1

:: Investment ::

The _____ is a measure of an investment's rate of return. The term internal refers to the fact that the calculation excludes external factors, such as the risk-free rate, inflation, the cost of capital, or various financial risks.

Exam Probability: **Medium**

28. *Answer choices:*

(see index for correct answer)

- a. Exchange fund
- b. Qirad
- c. Juniperus Capital
- d. With-profits policy

Guidance: level 1

:: Occupations ::

An _____ is a practitioner of accounting or accountancy, which is the measurement, disclosure or provision of assurance about financial information that helps managers, investors, tax authorities and others make decisions about allocating resource.

Exam Probability: **Medium**

29. *Answer choices:*

(see index for correct answer)

- a. Archivist
- b. Special Advocate
- c. Accountant
- d. Shopkeeper

Guidance: level 1

:: Project management ::

Some scenarios associate "this kind of planning" with learning "life skills". _____ s are necessary, or at least useful, in situations where individuals need to know what time they must be at a specific location to receive a specific service, and where people need to accomplish a set of goals within a set time period.

Exam Probability: **Low**

30. *Answer choices:*

(see index for correct answer)

- a. Cost estimate
- b. Legal matter management

- c. Schedule
- d. Duration

Guidance: level 1

:: ::

A _____ is the process of presenting a topic to an audience. It is typically a demonstration, introduction, lecture, or speech meant to inform, persuade, inspire, motivate, or to build good will or to present a new idea or product. The term can also be used for a formal or ritualized introduction or offering, as with the _____ of a debutante. _____ s in certain formats are also known as keynote address.

Exam Probability: **High**

31. *Answer choices:*

(see index for correct answer)

- a. Presentation
- b. imperative
- c. hierarchical perspective
- d. personal values

Guidance: level 1

:: Financial markets ::

A _____ is a market in which people trade financial securities and derivatives such as futures and options at low transaction costs. Securities include stocks and bonds, and precious metals.

Exam Probability: **High**

32. *Answer choices:*

(see index for correct answer)

- a. Financial market
- b. Broker-dealer
- c. Convenience yield
- d. Ugly Americans: The True Story of the Ivy League Cowboys Who Raided the Asian Markets for Millions

Guidance: level 1

:: Budgets ::

A _____ is a financial plan for a defined period, often one year. It may also include planned sales volumes and revenues, resource quantities, costs and expenses, assets, liabilities and cash flows. Companies, governments, families and other organizations use it to express strategic plans of activities or events in measurable terms.

Exam Probability: **Low**

33. *Answer choices:*

(see index for correct answer)

- a. Budget constraint
- b. Public budgeting
- c. Marginal budgeting for bottlenecks
- d. Budget

Guidance: level 1

:: Fraud ::

In law, _____ is intentional deception to secure unfair or unlawful gain, or to deprive a victim of a legal right. _____ can violate civil law, a criminal law, or it may cause no loss of money, property or legal right but still be an element of another civil or criminal wrong. The purpose of _____ may be monetary gain or other benefits, for example by obtaining a passport, travel document, or driver's license, or mortgage _____, where the perpetrator may attempt to qualify for a mortgage by way of false statements.

Exam Probability: **Medium**

34. *Answer choices:*

(see index for correct answer)

- a. Credit card kiting
- b. Insurance fraud

- c. Statute of frauds
- d. Tunneling

Guidance: level 1

:: Data analysis ::

In statistics, the _____ is a measure that is used to quantify the amount of variation or dispersion of a set of data values. A low _____ indicates that the data points tend to be close to the mean of the set, while a high _____ indicates that the data points are spread out over a wider range of values.

Exam Probability: **Low**

35. *Answer choices:*
(see index for correct answer)

- a. Probit
- b. Standard deviation
- c. Inverse-variance weighting
- d. Univariate analysis

Guidance: level 1

:: ::

_____ is the process of making predictions of the future based on past and present data and most commonly by analysis of trends. A commonplace example might be estimation of some variable of interest at some specified future date. Prediction is a similar, but more general term. Both might refer to formal statistical methods employing time series, cross-sectional or longitudinal data, or alternatively to less formal judgmental methods. Usage can differ between areas of application: for example, in hydrology the terms "forecast" and " _____ " are sometimes reserved for estimates of values at certain specific future times, while the term "prediction" is used for more general estimates, such as the number of times floods will occur over a long period.

Exam Probability: **Low**

36. *Answer choices:*

(see index for correct answer)

- a. levels of analysis
- b. Forecasting
- c. co-culture
- d. process perspective

Guidance: level 1

:: Money market instruments ::

_____ , in the global financial market, is an unsecured promissory note with a fixed maturity of not more than 270 days.

Exam Probability: **High**

37. *Answer choices:*

(see index for correct answer)

- a. Banker's acceptance
- b. Commercial paper

Guidance: level 1

:: Mereology ::

_____ , in the abstract, is what belongs to or with something, whether as an attribute or as a component of said thing. In the context of this article, it is one or more components , whether physical or incorporeal, of a person's estate; or so belonging to, as in being owned by, a person or jointly a group of people or a legal entity like a corporation or even a society. Depending on the nature of the _____ , an owner of _____ has the right to consume, alter, share, redefine, rent, mortgage, pawn, sell, exchange, transfer, give away or destroy it, or to exclude others from doing these things, as well as to perhaps abandon it; whereas regardless of the nature of the _____ , the owner thereof has the right to properly use it , or at the very least exclusively keep it.

Exam Probability: **High**

38. *Answer choices:*

(see index for correct answer)

- a. Mereology
- b. Property
- c. Meronomy
- d. Mereotopology

Guidance: level 1

:: Loans ::

In finance, a _____ is the lending of money by one or more individuals, organizations, or other entities to other individuals, organizations etc. The recipient incurs a debt, and is usually liable to pay interest on that debt until it is repaid, and also to repay the principal amount borrowed.

Exam Probability: **High**

39. *Answer choices:*
(see index for correct answer)

- a. Loan deficiency payments
- b. Fixed interest rate loan
- c. Package loan
- d. Term loan

Guidance: level 1

:: Accounting ::

It is the period for which books are balanced and the financial statements are prepared. Generally, the _____ consists of 12 months. However the beginning of the _____ differs according to the jurisdiction. For example, one entity may follow the regular calendar year, i.e. January to December as the accounting year, while another entity may follow April to March as the _____ .

Exam Probability: **Medium**

40. *Answer choices:*

(see index for correct answer)

- a. ACSOI
- b. FreeAgent
- c. Merdiban
- d. Accounting period

Guidance: level 1

:: Actuarial science ::

_____ services are provided by some large financial institutions, such as banks, or insurance or investment houses, whereby they guarantee payment in case of damage or financial loss and accept the financial risk for liability arising from such guarantee. An _____ arrangement may be created in a number of situations including insurance, issue of securities in a public offering, and bank lending, among others. The person or institution that agrees to sell a minimum number of securities of the company for commission is called the underwriter.

Exam Probability: **Medium**

41. *Answer choices:*

(see index for correct answer)

- a. Enterprise risk management
- b. General insurance
- c. Underwriting
- d. Value at risk

Guidance: level 1

:: Bonds (finance) ::

An _____ is a legal contract that reflects or covers a debt or purchase obligation. It specifically refers to two types of practices: in historical usage, an _____ d servant status, and in modern usage, it is an instrument used for commercial debt or real estate transaction.

Exam Probability: **High**

42. *Answer choices:*

(see index for correct answer)

- a. Mezzanine capital
- b. Panda bonds
- c. Indenture
- d. Agency debt

Guidance: level 1

:: Stock market ::

_____ is a form of stock which may have any combination of features not possessed by common stock including properties of both an equity and a debt instrument, and is generally considered a hybrid instrument. _____ s are senior to common stock, but subordinate to bonds in terms of claim and may have priority over common stock in the payment of dividends and upon liquidation. Terms of the _____ are described in the issuing company`s articles of association or articles of incorporation.

Exam Probability: **Low**

43. *Answer choices:*

(see index for correct answer)

- a. Wash sale

- b. Squeeze out
- c. Big boy letter
- d. Preferred stock

Guidance: level 1

:: ::

_____ , often abbreviated as B/E in finance, is the point of balance making neither a profit nor a loss. The term originates in finance but the concept has been applied in other fields.

Exam Probability: **Medium**

44. *Answer choices:*

(see index for correct answer)

- a. hierarchical
- b. similarity-attraction theory
- c. hierarchical perspective
- d. information systems assessment

Guidance: level 1

:: Capital gains taxes ::

A _____ refers to profit that results from a sale of a capital asset, such as stock, bond or real estate, where the sale price exceeds the purchase price. The gain is the difference between a higher selling price and a lower purchase price. Conversely, a capital loss arises if the proceeds from the sale of a capital asset are less than the purchase price.

Exam Probability: **Low**

45. *Answer choices:*

(see index for correct answer)

- a. Capital Cost Allowance
- b. Capital gain
- c. Capital cost tax factor

Guidance: level 1

:: Marketing ::

A _____ is the quantity of payment or compensation given by one party to another in return for one unit of goods or services.. A _____ is influenced by both production costs and demand for the product. A _____ may be determined by a monopolist or may be imposed on the firm by market conditions.

Exam Probability: **Low**

46. *Answer choices:*

(see index for correct answer)

- a. Free Comic Book Day
- b. Observatory of prices
- c. Price
- d. Product proliferation

Guidance: level 1

:: Expense ::

An _____ , operating expenditure, operational expense, operational expenditure or opex is an ongoing cost for running a product, business, or system. Its counterpart, a capital expenditure , is the cost of developing or providing non-consumable parts for the product or system. For example, the purchase of a photocopier involves capex, and the annual paper, toner, power and maintenance costs represents opex. For larger systems like businesses, opex may also include the cost of workers and facility expenses such as rent and utilities.

Exam Probability: **High**

47. *Answer choices:*

(see index for correct answer)

- a. expenditure
- b. Momentem

- c. Operating expense
- d. Stock option expensing

Guidance: level 1

:: Finance ::

The _____ of a corporation is the accumulated net income of the corporation that is retained by the corporation at a particular point of time, such as at the end of the reporting period. At the end of that period, the net income at that point is transferred from the Profit and Loss Account to the _____ account. If the balance of the _____ account is negative it may be called accumulated losses, retained losses or accumulated deficit, or similar terminology.

Exam Probability: **High**

48. *Answer choices:*

(see index for correct answer)

- a. Spot date
- b. CBDC NORTIP
- c. Unearned income
- d. Retained earnings

Guidance: level 1

:: Elementary geometry ::

The _____ is the front of an animal's head that features three of the head's sense organs, the eyes, nose, and mouth, and through which animals express many of their emotions. The _____ is crucial for human identity, and damage such as scarring or developmental deformities affects the psyche adversely.

Exam Probability: **High**

49. *Answer choices:*

(see index for correct answer)

- a. Central angle
- b. Median
- c. Hinge theorem
- d. Skew lines

Guidance: level 1

:: Management ::

The _____ is a strategy performance management tool – a semi-standard structured report, that can be used by managers to keep track of the execution of activities by the staff within their control and to monitor the consequences arising from these actions.

Exam Probability: **Medium**

50. *Answer choices:*

(see index for correct answer)

- a. Business relationship management
- b. Systems analysis
- c. Balanced scorecard
- d. Control

Guidance: level 1

:: Financial regulatory authorities of the United States ::

The _____ is the revenue service of the United States federal government. The government agency is a bureau of the Department of the Treasury, and is under the immediate direction of the Commissioner of Internal Revenue, who is appointed to a five-year term by the President of the United States. The IRS is responsible for collecting taxes and administering the Internal Revenue Code, the main body of federal statutory tax law of the United States. The duties of the IRS include providing tax assistance to taxpayers and pursuing and resolving instances of erroneous or fraudulent tax filings. The IRS has also overseen various benefits programs, and enforces portions of the Affordable Care Act.

Exam Probability: **Low**

51. *Answer choices:*

(see index for correct answer)

- a. Consumer Financial Protection Bureau
- b. Internal Revenue Service
- c. Farm Credit Administration
- d. Commodity Futures Trading Commission

Guidance: level 1

:: Financial markets ::

In economics and finance, _____ is the practice of taking advantage of a price difference between two or more markets: striking a combination of matching deals that capitalize upon the imbalance, the profit being the difference between the market prices. When used by academics, an _____ is a transaction that involves no negative cash flow at any probabilistic or temporal state and a positive cash flow in at least one state; in simple terms, it is the possibility of a risk-free profit after transaction costs. For example, an _____ opportunity is present when there is the opportunity to instantaneously buy something for a low price and sell it for a higher price.

Exam Probability: **Medium**

52. *Answer choices:*

(see index for correct answer)

- a. Spread trade
- b. Arbitrage
- c. Exchange of futures for swaps
- d. Post earnings announcement drift

Guidance: level 1

:: Financial risk ::

_____ is any of various types of risk associated with financing, including financial transactions that include company loans in risk of default. Often it is understood to include only downside risk, meaning the potential for financial loss and uncertainty about its extent.

Exam Probability: **Low**

53. *Answer choices:*

(see index for correct answer)

- a. Credit scorecards
- b. Time consistency
- c. Liquidity risk
- d. Cascades in Financial Networks

Guidance: level 1

:: ::

A _____ , or holiday, is a leave of absence from a regular occupation, or a specific trip or journey, usually for the purpose of recreation or tourism. People often take a _____ during specific holiday observances, or for specific festivals or celebrations. _____ s are often spent with friends or family.

Exam Probability: **Medium**

54. *Answer choices:*

(see index for correct answer)

- a. empathy
- b. interpersonal communication
- c. surface-level diversity
- d. hierarchical perspective

Guidance: level 1

:: Cash flow ::

_____ s are narrowly interconnected with the concepts of value, interest rate and liquidity. A _____ that shall happen on a future day tN can be transformed into a _____ of the same value in t0.

Exam Probability: **Medium**

55. *Answer choices:*

(see index for correct answer)

- a. Discounted cash flow
- b. Cash flow forecasting
- c. Cash flow
- d. Cash flow statement

Guidance: level 1

:: ::

An _____ is an asset that lacks physical substance. It is defined in opposition to physical assets such as machinery and buildings. An _____ is usually very hard to evaluate. Patents, copyrights, franchises, goodwill, trademarks, and trade names. The general interpretation also includes software and other intangible computer based assets are all examples of _____ s. _____ s generally—though not necessarily—suffer from typical market failures of non-rivalry and non-excludability.

Exam Probability: **Medium**

56. *Answer choices:*

(see index for correct answer)

- a. process perspective
- b. personal values
- c. imperative
- d. Intangible asset

Guidance: level 1

:: Bonds (finance) ::

A _____ is a fund established by an economic entity by setting aside revenue over a period of time to fund a future capital expense, or repayment of a long-term debt.

Exam Probability: **Medium**

57. *Answer choices:*
(see index for correct answer)

- a. Sinking fund
- b. Synthetic bond
- c. Collateralized debt obligation
- d. Luxembourg Depositary Receipt

Guidance: level 1

:: Business ::

The seller, or the provider of the goods or services, completes a sale in response to an acquisition, appropriation, requisition or a direct interaction with the buyer at the point of sale. There is a passing of title of the item, and the settlement of a price, in which agreement is reached on a price for which transfer of ownership of the item will occur. The seller, not the purchaser typically executes the sale and it may be completed prior to the obligation of payment. In the case of indirect interaction, a person who sells goods or service on behalf of the owner is known as a _____ man or _____ woman or _____ person, but this often refers to someone selling goods in a store/shop, in which case other terms are also common, including _____ clerk, shop assistant, and retail clerk.

Exam Probability: **High**

58. *Answer choices:*

(see index for correct answer)

- a. Disadvantaged business enterprise
- b. Business interoperability interface
- c. Sales
- d. E-lancing

Guidance: level 1

:: Actuarial science ::

The _____ is the greater benefit of receiving money now rather than an identical sum later. It is founded on time preference.

Exam Probability: **Medium**

59. *Answer choices:*

(see index for correct answer)

- a. Time value of money
- b. Future interests
- c. Demography
- d. Actuarial notation

Guidance: level 1

Human resource management

Human resource (HR) management is the strategic approach to the effective management of organization workers so that they help the business gain a competitive advantage. It is designed to maximize employee performance in service of an employer's strategic objectives. HR is primarily concerned with the management of people within organizations, focusing on policies and on systems. HR departments are responsible for overseeing employee-benefits design, employee recruitment, training and development, performance appraisal, and rewarding (e.g., managing pay and benefit systems). HR also concerns itself with organizational change and industrial relations, that is, the balancing of organizational practices with requirements arising from collective bargaining and from governmental laws.

:: Majority–minority relations ::

_____ , also known as reservation in India and Nepal, positive discrimination / action in the United Kingdom, and employment equity in Canada and South Africa, is the policy of promoting the education and employment of members of groups that are known to have previously suffered from discrimination. Historically and internationally, support for _____ has sought to achieve goals such as bridging inequalities in employment and pay, increasing access to education, promoting diversity, and redressing apparent past wrongs, harms, or hindrances.

Exam Probability: **High**

1. *Answer choices:*

(see index for correct answer)

- a. cultural Relativism
- b. Affirmative action
- c. positive discrimination

Guidance: level 1

:: Survey methodology ::

An _____ is a conversation where questions are asked and answers are given. In common parlance, the word " _____ " refers to a one-on-one conversation between an _____ er and an _____ ee. The _____ er asks questions to which the _____ ee responds, usually so information may be transferred from _____ ee to _____ er . Sometimes, information can be transferred in both directions. It is a communication, unlike a speech, which produces a one-way flow of information.

Exam Probability: **High**

2. *Answer choices:*

(see index for correct answer)

- a. Computer-assisted survey information collection
- b. National Health Interview Survey
- c. Survey sampling
- d. Interview

Guidance: level 1

:: Behaviorism ::

In behavioral psychology, _____ is a consequence applied that will strengthen an organism's future behavior whenever that behavior is preceded by a specific antecedent stimulus. This strengthening effect may be measured as a higher frequency of behavior, longer duration, greater magnitude, or shorter latency. There are two types of _____, known as positive _____ and negative _____; positive is where by a reward is offered on expression of the wanted behaviour and negative is taking away an undesirable element in the persons environment whenever the desired behaviour is achieved.

Exam Probability: **High**

3. *Answer choices:*

(see index for correct answer)

- a. chaining
- b. Matching Law
- c. Programmed instruction
- d. Reinforcement

Guidance: level 1

:: Educational assessment and evaluation ::

An _____ is a component of a competence to do a certain kind of work at a certain level. Outstanding _____ can be considered "talent". An _____ may be physical or mental. _____ is inborn potential to do certain kinds of work whether developed or undeveloped. Ability is developed knowledge, understanding, learned or acquired abilities or attitude. The innate nature of _____ is in contrast to skills and achievement, which represent knowledge or ability that is gained through learning.

Exam Probability: **Low**

4. *Answer choices:*

(see index for correct answer)

- a. Class rank
- b. Aptitude
- c. Achievement test
- d. Faculty Scholarly Productivity Index

Guidance: level 1

:: Options (finance) ::

_____ is a contractual agreement between a corporation and recipients of phantom shares that bestow upon the grantee the right to a cash payment at a designated time or in association with a designated event in the future, which payment is to be in an amount tied to the market value of an equivalent number of shares of the corporation's stock. Thus, the amount of the payout will increase as the stock price rises, and decrease if the stock falls, but without the recipient actually receiving any stock. Like other forms of stock-based compensation plans, _____ broadly serves to align the interests of recipients and shareholders, incent contribution to share value, and encourage the retention or continued participation of contributors. Recipients are typically employees, but may also be directors, third-party vendors, or others.

Exam Probability: **Medium**

5. *Answer choices:*
(see index for correct answer)

- a. LEAPS
- b. Greenspan put
- c. Phantom stock
- d. Contingent value rights

Guidance: level 1

:: Business terms ::

A _____ is a short statement of why an organization exists, what its overall goal is, identifying the goal of its operations: what kind of product or service it provides, its primary customers or market, and its geographical region of operation. It may include a short statement of such fundamental matters as the organization's values or philosophies, a business's main competitive advantages, or a desired future state—the "vision".

Exam Probability: **Medium**

6. *Answer choices:*

(see index for correct answer)

- a. Owner Controlled Insurance Program
- b. operating cost
- c. centralization
- d. Mission statement

Guidance: level 1

:: Employment compensation ::

A _____ is the minimum income necessary for a worker to meet their basic needs. Needs are defined to include food, housing, and other essential needs such as clothing. The goal of a _____ is to allow a worker to afford a basic but decent standard of living. Due to the flexible nature of the term "needs", there is not one universally accepted measure of what a _____ is and as such it varies by location and household type.

Exam Probability: **Low**

7. *Answer choices:*

(see index for correct answer)

- a. Living wage
- b. Basic income
- c. Golden handshake
- d. Golden parachute

Guidance: level 1

:: Recruitment ::

The _____ is an American nonprofit professional association established in 1956 in Bethlehem, Pennsylvania, for college career services, recruiting practitioners, and others who wish to hire the college educated.

Exam Probability: **Low**

8. *Answer choices:*

(see index for correct answer)

- a. South West African Native Labour Association
- b. Talent community
- c. National Association of Colleges and Employers
- d. Higher Education Recruitment Consortium

Guidance: level 1

:: Outsourcing ::

_____ is the relocation of a business process from one country to another—typically an operational process, such as manufacturing, or supporting processes, such as accounting. Typically this refers to a company business, although state governments may also employ _____ . More recently, technical and administrative services have been offshored.

Exam Probability: **Low**

9. *Answer choices:*

(see index for correct answer)

- a. Offshoring
- b. Virtual Staff Finder
- c. Cloud storage
- d. IQor

Guidance: level 1

:: ::

An _____ is a person temporarily or permanently residing in a country other than their native country. In common usage, the term often refers to professionals, skilled workers, or artists taking positions outside their home country, either independently or sent abroad by their employers, who can be companies, universities, governments, or non-governmental organisations. Effectively migrant workers, they usually earn more than they would at home, and less than local employees. However, the term ` _____ ` is also used for retirees and others who have chosen to live outside their native country. Historically, it has also referred to exiles.

Exam Probability: **Medium**

10. *Answer choices:*

(see index for correct answer)

- a. levels of analysis
- b. Expatriate
- c. open system
- d. similarity-attraction theory

Guidance: level 1

:: Human resource management ::

_____ or work sharing is an employment arrangement where typically two people are retained on a part-time or reduced-time basis to perform a job normally fulfilled by one person working full-time. Since all positions are shared thus leads to a net reduction in per-employee income. The people sharing the job work as a team to complete the job task and are equally responsible for the job workload. Compensation is apportioned between the workers. Working hours, pay and holidays are divided equally. The pay as you go system helps make deductions for national insurance and superannuations are made as a straightforward percentage.

Exam Probability: **Low**

11. *Answer choices:*

(see index for correct answer)

- a. Job sharing
- b. Lego Serious Play
- c. Succession planning
- d. Leadership development

Guidance: level 1

:: Employment compensation ::

_____ is a notional derivative of a Health Reimbursement Arrangement, a type of US employer-funded health benefit plan that reimburses employees for out-of-pocket medical expenses and, in limited cases, to pay for health insurance plan premiums.

Exam Probability: **Medium**

12. *Answer choices:*

(see index for correct answer)

- a. Non-wage labour costs
- b. Health Reimbursement Account
- c. Merit pay
- d. Employee assistance program

Guidance: level 1

:: Problem solving ::

A _____ is a unit or formation established to work on a single defined task or activity. Originally introduced by the United States Navy, the term has now caught on for general usage and is a standard part of NATO terminology. Many non-military organizations now create " _____ s" or task groups for temporary activities that might have once been performed by ad hoc committees.

Exam Probability: **Medium**

13. *Answer choices:*

(see index for correct answer)

- a. Circle time
- b. Cognitive acceleration

- c. Task force
- d. Puzzle

Guidance: level 1

:: Employment compensation ::

Compensation and benefits is a sub-discipline of human resources, focused on employee compensation and benefits policy-making. While compensation and benefits are tangible, there are intangible rewards such as recognition, work-life and development. Combined, these are referred to as _____ s . The term "compensation and benefits" refers to the discipline as well as the rewards themselves.

Exam Probability: **Low**

14. *Answer choices:*

(see index for correct answer)

- a. Basic income
- b. Merit pay
- c. Prevailing wage
- d. Total Reward

Guidance: level 1

:: Employment discrimination ::

A _____ is a metaphor used to represent an invisible barrier that keeps a given demographic from rising beyond a certain level in a hierarchy.

Exam Probability: **Low**

15. *Answer choices:*

(see index for correct answer)

- a. Glass ceiling
- b. New South Wales selection bias
- c. Employment discrimination law in the European Union
- d. LGBT employment discrimination in the United States

Guidance: level 1

:: Unemployment benefits ::

_____ are payments made by back authorized bodies to unemployed people. In the United States, benefits are funded by a compulsory governmental insurance system, not taxes on individual citizens. Depending on the jurisdiction and the status of the person, those sums may be small, covering only basic needs, or may compensate the lost time proportionally to the previous earned salary.

Exam Probability: **Low**

16. *Answer choices:*

(see index for correct answer)

- a. Unemployment benefits in Sweden
- b. National Insurance Act 1911
- c. Unemployment benefits in Spain
- d. Unemployment benefits

Guidance: level 1

:: Labor terms ::

_____ , often called DI or disability income insurance, or income protection, is a form of insurance that insures the beneficiary's earned income against the risk that a disability creates a barrier for a worker to complete the core functions of their work. For example, the worker may suffer from an inability to maintain composure in the case of psychological disorders or an injury, illness or condition that causes physical impairment or incapacity to work. It encompasses paid sick leave, short-term disability benefits , and long-term disability benefits . Statistics show that in the US a disabling accident occurs, on average, once every second. In fact, nearly 18.5% of Americans are currently living with a disability, and 1 out of every 4 persons in the US workforce will suffer a disabling injury before retirement.

Exam Probability: **High**

17. *Answer choices:*

(see index for correct answer)

- a. Civilian workers

- b. Absence rate
- c. Capital services
- d. Consumer unit

Guidance: level 1

:: Labor ::

The workforce or labour force is the labour pool in employment. It is generally used to describe those working for a single company or industry, but can also apply to a geographic region like a city, state, or country. Within a company, its value can be labelled as its "Workforce in Place". The workforce of a country includes both the employed and the unemployed. The labour force participation rate, LFPR , is the ratio between the labour force and the overall size of their cohort . The term generally excludes the employers or management, and can imply those involved in manual labour. It may also mean all those who are available for work.

Exam Probability: **Medium**

18. *Answer choices:*

(see index for correct answer)

- a. Labor force
- b. Man-hour
- c. Departmentalization
- d. Quality of working life

Guidance: level 1

:: ::

A _____ is a research instrument consisting of a series of questions for the purpose of gathering information from respondents. The _____ was invented by the Statistical Society of London in 1838.

Exam Probability: **Medium**

19. *Answer choices:*

(see index for correct answer)

- a. imperative
- b. empathy
- c. process perspective
- d. Questionnaire

Guidance: level 1

:: Financial statements ::

In financial accounting, a _____ or statement of financial position or statement of financial condition is a summary of the financial balances of an individual or organization, whether it be a sole proprietorship, a business partnership, a corporation, private limited company or other organization such as Government or not-for-profit entity. Assets, liabilities and ownership equity are listed as of a specific date, such as the end of its financial year. A _____ is often described as a "snapshot of a company's financial condition". Of the four basic financial statements, the _____ is the only statement which applies to a single point in time of a business' calendar year.

Exam Probability: **High**

20. *Answer choices:*

(see index for correct answer)

- a. Government financial statements
- b. Financial report
- c. Balance sheet
- d. Clean surplus accounting

Guidance: level 1

:: Cognitive biases ::

The _____ is a type of immediate judgement discrepancy, or cognitive bias, where a person making an initial assessment of another person, place, or thing will assume ambiguous information based upon concrete information. A simplified example of the _____ is when an individual noticing that the person in the photograph is attractive, well groomed, and properly attired, assumes, using a mental heuristic, that the person in the photograph is a good person based upon the rules of that individual's social concept. This constant error in judgment is reflective of the individual's preferences, prejudices, ideology, aspirations, and social perception. The _____ is an evaluation by an individual and can affect the perception of a decision, action, idea, business, person, group, entity, or other whenever concrete data is generalized or influences ambiguous information.

Exam Probability: **High**

21. *Answer choices:*

(see index for correct answer)

- a. Fundamental attribution error
- b. Moral credential
- c. Straight and Crooked Thinking
- d. Halo effect

Guidance: level 1

:: ::

From an accounting perspective, _____ is crucial because _____ and _____ taxes considerably affect the net income of most companies and because they are subject to laws and regulations.

Exam Probability: **High**

22. *Answer choices:*

(see index for correct answer)

- a. Payroll
- b. Character
- c. personal values
- d. surface-level diversity

Guidance: level 1

:: Corporate governance ::

An _____ is generally a person responsible for running an organization, although the exact nature of the role varies depending on the organization. In many militaries, an _____ , or "XO," is the second-in-command, reporting to the commanding officer. The XO is typically responsible for the management of day-to-day activities, freeing the commander to concentrate on strategy and planning the unit's next move.

Exam Probability: **High**

23. *Answer choices:*

(see index for correct answer)

- a. Directors and officers liability insurance
- b. Chief analytics officer
- c. Short swing
- d. Chief operating officer

Guidance: level 1

:: Industrial engineering ::

_____ is the formal process that sits alongside Requirements analysis and focuses on the human elements of the requirements.

Exam Probability: **High**

24. *Answer choices:*

(see index for correct answer)

- a. Institute of Industrial Engineers
- b. Needs analysis
- c. Work sampling
- d. Worker-machine activity chart

Guidance: level 1

:: Business law ::

An _____ is a natural person, business, or corporation that provides goods or services to another entity under terms specified in a contract or within a verbal agreement. Unlike an employee, an _____ does not work regularly for an employer but works as and when required, during which time they may be subject to law of agency. _____ s are usually paid on a freelance basis. Contractors often work through a limited company or franchise, which they themselves own, or may work through an umbrella company.

Exam Probability: **High**

25. *Answer choices:*
(see index for correct answer)

- a. Independent contractor
- b. Forged endorsement
- c. Complex structured finance transactions
- d. Business.gov

Guidance: level 1

:: Self ::

_____ is a term that has been used in various psychology theories, often in different ways. The term was originally introduced by the organismic theorist Kurt Goldstein for the motive to realize one's full potential. In Goldstein's view, it is the organism's master motive, the only real motive: "the tendency to actualize itself as fully as possible is the basic drive ... the drive of _____ ." Carl Rogers similarly wrote of "the curative force in psychotherapy man's tendency to actualize himself, to become his potentialities ... to express and activate all the capacities of the organism." The concept was brought most fully to prominence in Abraham Maslow's hierarchy of needs theory as the final level of psychological development that can be achieved when all basic and mental needs are essentially fulfilled and the "actualization" of the full personal potential takes place, although he adapted this viewpoint later on in life to be more flexible.

Exam Probability: **Low**

26. *Answer choices:*

(see index for correct answer)

- a. Self-presentation
- b. ecological self
- c. Narcissism
- d. Self-actualization

Guidance: level 1

:: Employment ::

_____ is a relationship between two parties, usually based on a contract where work is paid for, where one party, which may be a corporation, for profit, not-for-profit organization, co-operative or other entity is the employer and the other is the employee. Employees work in return for payment, which may be in the form of an hourly wage, by piecework or an annual salary, depending on the type of work an employee does or which sector she or he is working in. Employees in some fields or sectors may receive gratuities, bonus payment or stock options. In some types of _____ , employees may receive benefits in addition to payment. Benefits can include health insurance, housing, disability insurance or use of a gym. _____ is typically governed by _____ laws, regulations or legal contracts.

Exam Probability: **Medium**

27. *Answer choices:*

(see index for correct answer)

- a. Employment
- b. Job Services Australia
- c. Blue-collar worker
- d. Job attitude

Guidance: level 1

:: Management ::

_____ is a set of activities that ensure goals are met in an effective and efficient manner. _____ can focus on the performance of an organization, a department, an employee, or the processes in place to manage particular tasks. _____ standards are generally organized and disseminated by senior leadership at an organization, and by task owners.

Exam Probability: **Medium**

28. *Answer choices:*

(see index for correct answer)

- a. Balanced scorecard
- b. Supply chain sustainability
- c. Performance management
- d. Top development

Guidance: level 1

:: Employment compensation ::

Employee stock ownership, or employee share ownership, is an ownership interest in a company held by the company's workforce. The ownership interest may be facilitated by the company as part of employees' remuneration or incentive compensation for work performed, or the company itself may be employee owned.

Exam Probability: **Low**

29. *Answer choices:*

(see index for correct answer)

- a. Pension insurance contract
- b. Employee stock ownership plan
- c. Long service leave
- d. Employee benefit

Guidance: level 1

:: Minimum wage ::

A _____ is the lowest remuneration that employers can legally pay their workers—the price floor below which workers may not sell their labor. Most countries had introduced _____ legislation by the end of the 20th century.

Exam Probability: **High**

30. *Answer choices:*

(see index for correct answer)

- a. Minimum wage in the United States
- b. Minimum wage
- c. Minimum Wage Fairness Act
- d. Working poor

Guidance: level 1

:: Recruitment ::

A _____ is a quantitative research method commonly employed in survey research. The aim of this approach is to ensure that each interview is presented with exactly the same questions in the same order. This ensures that answers can be reliably aggregated and that comparisons can be made with confidence between sample subgroups or between different survey periods.

Exam Probability: **High**

31. *Answer choices:*

(see index for correct answer)

- a. Acqui-hiring
- b. Structured interview
- c. Online job fair
- d. Military recruitment

Guidance: level 1

:: ::

A _____, medical practitioner, medical doctor, or simply doctor, is a professional who practises medicine, which is concerned with promoting, maintaining, or restoring health through the study, diagnosis, prognosis and treatment of disease, injury, and other physical and mental impairments. _____s may focus their practice on certain disease categories, types of patients, and methods of treatment—known as specialities—or they may assume responsibility for the provision of continuing and comprehensive medical care to individuals, families, and communities—known as general practice. Medical practice properly requires both a detailed knowledge of the academic disciplines, such as anatomy and physiology, underlying diseases and their treatment—the science of medicine—and also a decent competence in its applied practice—the art or craft of medicine.

Exam Probability: **Low**

32. *Answer choices:*

(see index for correct answer)

- a. hierarchical perspective
- b. empathy
- c. similarity-attraction theory
- d. surface-level diversity

Guidance: level 1

In production, research, retail, and accounting, a _____ is the value of money that has been used up to produce something or deliver a service, and hence is not available for use anymore. In business, the _____ may be one of acquisition, in which case the amount of money expended to acquire it is counted as _____ . In this case, money is the input that is gone in order to acquire the thing. This acquisition _____ may be the sum of the _____ of production as incurred by the original producer, and further _____ s of transaction as incurred by the acquirer over and above the price paid to the producer. Usually, the price also includes a mark-up for profit over the _____ of production.

Exam Probability: **Low**

33. *Answer choices:*

(see index for correct answer)

- a. hierarchical
- b. empathy
- c. Cost
- d. hierarchical perspective

Guidance: level 1

:: Social psychology ::

In social psychology, _____ is the phenomenon of a person exerting less effort to achieve a goal when he or she works in a group than when working alone. This is seen as one of the main reasons groups are sometimes less productive than the combined performance of their members working as individuals, but should be distinguished from the accidental coordination problems that groups sometimes experience.

Exam Probability: **Medium**

34. *Answer choices:*
(see index for correct answer)

- a. thought control
- b. Personal space
- c. Social loafing
- d. Mutual engagement

Guidance: level 1

:: Human resource management ::

_____ are the people who make up the workforce of an organization, business sector, or economy. "Human capital" is sometimes used synonymously with "_____", although human capital typically refers to a narrower effect. Likewise, other terms sometimes used include manpower, talent, labor, personnel, or simply people.

Exam Probability: **High**

35. *Answer choices:*

(see index for correct answer)

- a. Organizational culture
- b. Human resources
- c. Employee value proposition
- d. Service record

Guidance: level 1

:: ::

A _____ is a systematic way of determining the value/worth of a job in relation to other jobs in an organization. It tries to make a systematic comparison between jobs to assess their relative worth for the purpose of establishing a rational pay structure. _____ needs to be differentiated from job analysis. Job analysis is a systematic way of gathering information about a job. Every _____ method requires at least some basic job analysis in order to provide factual information about the jobs concerned. Thus, _____ begins with job analysis and ends at that point where the worth of a job is ascertained for achieving pay equity between jobs and different roles.

Exam Probability: **Medium**

36. *Answer choices:*

(see index for correct answer)

- a. Sarbanes-Oxley act of 2002
- b. similarity-attraction theory

- c. co-culture
- d. Job evaluation

Guidance: level 1

:: Stress ::

_____ means beneficial stress—either psychological, physical, or biochemical/radiological .

Exam Probability: **High**

37. *Answer choices:*

(see index for correct answer)

- a. Epinephrine
- b. Mental distress
- c. Roseto effect
- d. Eustress

Guidance: level 1

:: Trade unions ::

A _____ , in North America, or union branch, in the United Kingdom and other countries, is a local branch of a usually national trade union. The terms used for sub-branches of _____ s vary from country to country and include "shop committee", "shop floor committee", "board of control", "chapel", and others.

Exam Probability: **Medium**

38. *Answer choices:*

(see index for correct answer)

- a. National trade union center
- b. Global Labour University
- c. General union
- d. Service model

Guidance: level 1

:: Human resource management ::

_____ is a sub-discipline of human resources, focused on employee _____ policy-making. While _____ are tangible, there are intangible rewards such as recognition, work-life and development. Combined, these are referred to as total rewards . The term " _____ " refers to the discipline as well as the rewards themselves.

Exam Probability: **Medium**

39. *Answer choices:*

(see index for correct answer)

- a. Employee value proposition
- b. Compensation and benefits
- c. Management due diligence
- d. T-shaped skills

Guidance: level 1

:: United Kingdom labour law ::

The _____ was a series of programs, public work projects, financial reforms, and regulations enacted by President Franklin D. Roosevelt in the United States between 1933 and 1936. It responded to needs for relief, reform, and recovery from the Great Depression. Major federal programs included the Civilian Conservation Corps , the Civil Works Administration , the Farm Security Administration , the National Industrial Recovery Act of 1933 and the Social Security Administration . They provided support for farmers, the unemployed, youth and the elderly. The _____ included new constraints and safeguards on the banking industry and efforts to re-inflate the economy after prices had fallen sharply. _____ programs included both laws passed by Congress as well as presidential executive orders during the first term of the presidency of Franklin D. Roosevelt.

Exam Probability: **Low**

40. *Answer choices:*

(see index for correct answer)

- a. Working Time Regulations 1998
- b. Employers and Workmen Act 1875
- c. Labour Exchanges Act 1909
- d. Mutual trust and confidence

Guidance: level 1

:: ::

_____ is a method for employees to organize into a labor union in which a majority of employees in a bargaining unit sign authorization forms, or "cards", stating they wish to be represented by the union. Since the National Labor Relations Act became law in 1935, _____ has been an alternative to the National Labor Relations Board's election process. _____ and election are both overseen by the National Labor Relations Board. The difference is that with card sign-up, employees sign authorization cards stating they want a union, the cards are submitted to the NLRB and if more than 50% of the employees submitted cards, the NLRB requires the employer to recognize the union. The NLRA election process is an additional step with the NLRB conducting a secret ballot election after authorization cards are submitted. In both cases the employer never sees the authorization cards or any information that would disclose how individual employees voted.

Exam Probability: **Medium**

41. *Answer choices:*

(see index for correct answer)

- a. hierarchical
- b. functional perspective

- c. surface-level diversity
- d. empathy

Guidance: level 1

:: Business law ::

In professional sports, a _____ is a player who is eligible to freely sign with any club or franchise; i.e., not under contract to any specific team. The term is also used in reference to a player who is under contract at present but who is allowed to solicit offers from other teams. In some circumstances, the _____ 's options are limited by league rules.

Exam Probability: **High**

42. *Answer choices:*

(see index for correct answer)

- a. Trading while insolvent
- b. Financial Security Law of France
- c. Free agent
- d. Unfair preference

Guidance: level 1

:: ::

_____ is an important topic of Human Resource Management. It helps develop the career of the individual and the prosperous growth of the organization. On the job training is a form of training provided at the workplace. During the training, employees are familiarized with the working environment they will become part of. Employees also get a hands-on experience using machinery, equipment, tools, materials, etc. Part of is to face the challenges that occur during the performance of the job. An experienced employee or a manager are executing the role of the mentor who through written, or verbal instructions and demonstrations are passing on his/her knowledge and company-specific skills to the new employee. Executing the training on at the job location, rather than the classroom, creates a stress-free environment for the employees. _____ is the most popular method of training not only in the United States but in most of the developed countries, such as the United Kingdom, China, Russia, etc. Its effectiveness is based on the use of existing workplace tools, machines, documents and equipment, and the knowledge of specialists who are working in this field. _____ is easy to arrange and manage and it simplifies the process of adapting to the new workplace. OJT is highly used for practical tasks. It is inexpensive, and it doesn't require special equipment that is normally used for a specific job. Upon satisfaction of completion of the training, the employer is expected to retain participants as regular employees.

Exam Probability: **High**

43. *Answer choices:*

(see index for correct answer)

- a. On-the-job training
- b. corporate values
- c. imperative
- d. surface-level diversity

Guidance: level 1

:: Industrial relations ::

_____ or employee satisfaction is a measure of workers' contentedness with their job, whether or not they like the job or individual aspects or facets of jobs, such as nature of work or supervision. _____ can be measured in cognitive, affective, and behavioral components. Researchers have also noted that _____ measures vary in the extent to which they measure feelings about the job, or cognitions about the job.

Exam Probability: **High**

44. *Answer choices:*

(see index for correct answer)

- a. Workforce Investment Board
- b. Job satisfaction
- c. Industrial violence
- d. European Journal of Industrial Relations

Guidance: level 1

:: Workplace ::

_____ or occupational violence refers to violence, usually in the form of physical abuse or threat, that creates a risk to the health and safety of an employee or multiple employees. The National Institute for Occupational Safety and Health defines worker on worker, personal relationship, customer/client, and criminal intent all as categories of violence in the workplace. These four categories are further broken down into three levels: Level one displays early warning signs of violence, Level two is slightly more violent, and level three is significantly violent. Many workplaces have initiated programs and protocols to protect their workers as the Occupational Health Act of 1970 states that employers must provide an environment in which employees are free of harm or harmful conditions.

Exam Probability: **High**

45. *Answer choices:*

(see index for correct answer)

- a. Workplace aggression
- b. performance review
- c. Workplace violence
- d. Counterproductive work behavior

Guidance: level 1

_____ refers to the overall process of attracting, shortlisting, selecting and appointing suitable candidates for jobs within an organization. _____ can also refer to processes involved in choosing individuals for unpaid roles. Managers, human resource generalists and _____ specialists may be tasked with carrying out _____ , but in some cases public-sector employment agencies, commercial _____ agencies, or specialist search consultancies are used to undertake parts of the process. Internet-based technologies which support all aspects of _____ have become widespread.

Exam Probability: **High**

46. *Answer choices:*

(see index for correct answer)

- a. hierarchical perspective
- b. corporate values
- c. levels of analysis
- d. interpersonal communication

Guidance: level 1

:: Management education ::

_____ refers to simulation games that are used as an educational tool for teaching business. _____ s may be carried out for various business training such as: general management, finance, organizational behaviour, human resources, etc. Often, the term "business simulation" is used with the same meaning.

Exam Probability: **High**

47. *Answer choices:*

(see index for correct answer)

- a. Human systems engineering
- b. Academy of Management
- c. Business game
- d. Entrepreneurship education

Guidance: level 1

:: ::

A _____ is an occupation founded upon specialized educational training, the purpose of which is to supply disinterested objective counsel and service to others, for a direct and definite compensation, wholly apart from expectation of other business gain. The term is a truncation of the term "liberal _____ ", which is, in turn, an Anglicization of the French term " _____ libérale". Originally borrowed by English users in the 19th century, it has been re-borrowed by international users from the late 20th, though the class overtones of the term do not seem to survive retranslation: "liberal _____ s" are, according to the European Union's Directive on Recognition of _____ al Qualifications "those practiced on the basis of relevant _____ al qualifications in a personal, responsible and _____ ally independent capacity by those providing intellectual and conceptual services in the interest of the client and the public".

Exam Probability: **High**

48. Answer choices:

(see index for correct answer)

- a. deep-level diversity
- b. surface-level diversity
- c. personal values
- d. Sarbanes-Oxley act of 2002

Guidance: level 1

:: Validity (statistics) ::

In psychometrics, _____ is the extent to which a score on a scale or test predicts scores on some criterion measure.

Exam Probability: **Medium**

49. Answer choices:

(see index for correct answer)

- a. Face validity
- b. Construct validity
- c. Internal validity
- d. Predictive validity

Guidance: level 1

:: Management ::

> The _____ is a strategy performance management tool – a semi-standard structured report, that can be used by managers to keep track of the execution of activities by the staff within their control and to monitor the consequences arising from these actions.

Exam Probability: **High**

50. *Answer choices:*

(see index for correct answer)

- a. Stovepipe
- b. Balanced scorecard
- c. Director
- d. Vasa syndrome

Guidance: level 1

:: Financial terminology ::

_____ is the cost of maintaining a certain standard of living. Changes in the _____ over time are often operationalized in a cost-of-living index. _____ calculations are also used to compare the cost of maintaining a certain standard of living in different geographic areas. Differences in _____ between locations can also be measured in terms of purchasing power parity rates.

Exam Probability: **Medium**

51. *Answer choices:*

(see index for correct answer)

- a. Cost of living
- b. Multi-currency pricing
- c. Sale-and-leaseback
- d. Backwardation

Guidance: level 1

:: Labour law ::

In law, _____ is to give an immediately secured right of present or future deployment. One has a vested right to an asset that cannot be taken away by any third party, even though one may not yet possess the asset. When the right, interest, or title to the present or future possession of a legal estate can be transferred to any other party, it is termed a vested interest.

Exam Probability: **High**

52. Answer choices:

(see index for correct answer)

- a. Negligence in employment
- b. Collective agreement
- c. Works council
- d. Positive action

Guidance: level 1

:: Human resource management ::

A _____ is an outsourcing firm which provides services to small and medium sized businesses. Typically, the PEO offering may include human resource consulting, safety and risk mitigation services, payroll processing, employer payroll tax filing, workers' compensation insurance, health benefits, employers' practice and liability insurance, retirement vehicles, regulatory compliance assistance, workforce management technology, and training and development. The PEO enters into a contractual co-employment agreement with its clientele. Through co-employment, the PEO becomes the employer of record for tax purposes through filing payroll taxes under its own tax identification numbers.

Exam Probability: **Medium**

53. Answer choices:

(see index for correct answer)

- a. Professional employer organization

- b. Organizational chart
- c. Organizational culture
- d. Employee retention

Guidance: level 1

:: Management ::

_____ is a technique used by some employers to rotate their employees' assigned jobs throughout their employment. Employers practice this technique for a number of reasons. It was designed to promote flexibility of employees and to keep employees interested into staying with the company/organization which employs them. There is also research that shows how _____ s help relieve the stress of employees who work in a job that requires manual labor.

Exam Probability: **High**

54. *Answer choices:*
(see index for correct answer)

- a. Virtual customer environment
- b. Job rotation
- c. Managing stage boundaries
- d. Distributed management

Guidance: level 1

:: Employment ::

A flat organization has an organizational structure with few or no levels of middle management between staff and executives. An organization's structure refers to the nature of the distribution of the units and positions within it, also to the nature of the relationships among those units and positions. Tall and flat organizations differ based on how many levels of management are present in the organization, and how much control managers are endowed with.

Exam Probability: **High**

55. *Answer choices:*

(see index for correct answer)

- a. Blue-collar worker
- b. CIETC
- c. Hourly worker
- d. Delayering

Guidance: level 1

:: Leadership ::

_____ is a theory of leadership where a leader works with teams to identify needed change, creating a vision to guide the change through inspiration, and executing the change in tandem with committed members of a group; it is an integral part of the Full Range Leadership Model. _____ serves to enhance the motivation, morale, and job performance of followers through a variety of mechanisms; these include connecting the follower's sense of identity and self to a project and to the collective identity of the organization; being a role model for followers in order to inspire them and to raise their interest in the project; challenging followers to take greater ownership for their work, and understanding the strengths and weaknesses of followers, allowing the leader to align followers with tasks that enhance their performance.

Exam Probability: **Low**

56. *Answer choices:*

(see index for correct answer)

- a. servant leader
- b. BTS Group
- c. Three levels of leadership model
- d. European Center for Leadership Development

Guidance: level 1

:: ::

An _____ is a period of work experience offered by an organization for a limited period of time. Once confined to medical graduates, the term is now used for a wide range of placements in businesses, non-profit organizations and government agencies. They are typically undertaken by students and graduates looking to gain relevant skills and experience in a particular field. Employers benefit from these placements because they often recruit employees from their best interns, who have known capabilities, thus saving time and money in the long run. _____s are usually arranged by third-party organizations which recruit interns on behalf of industry groups. Rules vary from country to country about when interns should be regarded as employees. The system can be open to exploitation by unscrupulous employers.

Exam Probability: **Medium**

57. *Answer choices:*

(see index for correct answer)

- a. process perspective
- b. Internship
- c. Sarbanes-Oxley act of 2002
- d. surface-level diversity

Guidance: level 1

:: Labor ::

_____s are workers whose main capital is knowledge. Examples include programmers, physicians, pharmacists, architects, engineers, scientists, design thinkers, public accountants, lawyers, and academics, and any other white-collar workers, whose line of work requires the one to "think for a living".

Exam Probability: **High**

58. *Answer choices:*

(see index for correct answer)

- a. Knowledge worker
- b. Andrew Davison
- c. Designated Suppliers Program
- d. Departmentalization

Guidance: level 1

:: Self ::

_____ is a conscious or subconscious process in which people attempt to influence the perceptions of other people about a person, object or event. They do so by regulating and controlling information in social interaction. It was first conceptualized by Erving Goffman in 1959 in The Presentation of Self in Everyday Life, and then was expanded upon in 1967. An example of _____ theory in play is in sports such as soccer. At an important game, a player would want to showcase themselves in the best light possible, because there are college recruiters watching. This person would have the flashiest pair of cleats and try and perform their best to show off their skills. Their main goal may be to impress the college recruiters in a way that maximizes their chances of being chosen for a college team rather than winning the game.

Exam Probability: **High**

59. *Answer choices:*
(see index for correct answer)

- a. Impression management
- b. a person
- c. Self-actualization
- d. ecological self

Guidance: level 1

Information systems

Information systems (IS) are formal, sociotechnical, organizational systems designed to collect, process, store, and distribute information. In a sociotechnical perspective Information Systems are composed by four components: technology, process, people and organizational structure.

:: Credit cards ::

The _____ Company, also known as Amex, is an American multinational financial services corporation headquartered in Three World Financial Center in New York City. The company was founded in 1850 and is one of the 30 components of the Dow Jones Industrial Average. The company is best known for its charge card, credit card, and traveler's cheque businesses.

Exam Probability: **Medium**

1. *Answer choices:*

(see index for correct answer)

- a. BC Card
- b. Credit Saison
- c. Access
- d. American Express

Guidance: level 1

:: Automatic identification and data capture ::

_____ is human–computer interaction in which a computer is expected to be transported during normal usage, which allows for transmission of data, voice and video. _____ involves mobile communication, mobile hardware, and mobile software. Communication issues include ad hoc networks and infrastructure networks as well as communication properties, protocols, data formats and concrete technologies. Hardware includes mobile devices or device components. Mobile software deals with the characteristics and requirements of mobile applications.

Exam Probability: **Low**

2. *Answer choices:*

(see index for correct answer)

- a. CyberCode
- b. Molecular computational identification
- c. SILVIA
- d. Mobile computing

Guidance: level 1

:: Internet advertising ::

_____ is software that aims to gather information about a person or organization, sometimes without their knowledge, that may send such information to another entity without the consumer's consent, that asserts control over a device without the consumer's knowledge, or it may send such information to another entity with the consumer's consent, through cookies.

Exam Probability: **High**

3. *Answer choices:*

(see index for correct answer)

- a. Phorm
- b. Domain drop catching
- c. Comparison of ad servers
- d. Spyware

Guidance: level 1

:: Virtual economies ::

_____ Inc. is an American social game developer running social video game services founded in April 2007 and headquartered in San Francisco, California, United States. The company primarily focuses on mobile and social networking platforms. _____ states its mission as "connecting the world through games."

Exam Probability: **Medium**

4. *Answer choices:*

(see index for correct answer)

- a. Miniconomy
- b. Spiral Knights
- c. Zynga
- d. Pioneer

Guidance: level 1

:: Google services ::

_____ is a discontinued image organizer and image viewer for organizing and editing digital photos, plus an integrated photo-sharing website, originally created by a company named Lifescape in 2002. In July 2004, Google acquired _____ from Lifescape and began offering it as freeware. "_____" is a blend of the name of Spanish painter Pablo Picasso, the phrase mi casa and "pic" for pictures.

Exam Probability: **Medium**

5. *Answer choices:*

(see index for correct answer)

- a. Google App Engine
- b. Google Flights
- c. Google Scholar
- d. Google Translator Toolkit

Guidance: level 1

:: Economic globalization ::

_____ is an agreement in which one company hires another company to be responsible for a planned or existing activity that is or could be done internally, and sometimes involves transferring employees and assets from one firm to another.

Exam Probability: **Low**

6. *Answer choices:*

(see index for correct answer)

- a. reshoring
- b. Outsourcing

Guidance: level 1

:: Computing output devices ::

An _____ is any piece of computer hardware equipment which converts information into human-readable form.

Exam Probability: **Medium**

7. *Answer choices:*

(see index for correct answer)

- a. MyVu
- b. Indexed color
- c. DR37-P
- d. Output device

Guidance: level 1

:: User interfaces ::

The _____ , in the industrial design field of human–computer interaction, is the space where interactions between humans and machines occur. The goal of this interaction is to allow effective operation and control of the machine from the human end, whilst the machine simultaneously feeds back information that aids the operators' decision-making process. Examples of this broad concept of _____ s include the interactive aspects of computer operating systems, hand tools, heavy machinery operator controls, and process controls. The design considerations applicable when creating _____ s are related to or involve such disciplines as ergonomics and psychology.

Exam Probability: **High**

8. *Answer choices:*

(see index for correct answer)

- a. Baifox
- b. Principles of attention stress
- c. Browser user interface
- d. Monome

Guidance: level 1

:: Survey methodology ::

A _____ is the procedure of systematically acquiring and recording information about the members of a given population. The term is used mostly in connection with national population and housing _____ es; other common _____ es include agriculture, business, and traffic _____ es. The United Nations defines the essential features of population and housing _____ es as "individual enumeration, universality within a defined territory, simultaneity and defined periodicity", and recommends that population _____ es be taken at least every 10 years. United Nations recommendations also cover _____ topics to be collected, official definitions, classifications and other useful information to co-ordinate international practice.

Exam Probability: **Low**

9. *Answer choices:*

(see index for correct answer)

- a. Political forecasting
- b. Census
- c. Self-report
- d. National Health Interview Survey

Guidance: level 1

:: Internet privacy ::

An _____ is a private network accessible only to an organization's staff. Often, a wide range of information and services are available on an organization's internal _____ that are unavailable to the public, unlike the Internet. A company-wide _____ can constitute an important focal point of internal communication and collaboration, and provide a single starting point to access internal and external resources. In its simplest form, an _____ is established with the technologies for local area networks and wide area networks. Many modern _____ s have search engines, user profiles, blogs, mobile apps with notifications, and events planning within their infrastructure.

Exam Probability: **Low**

10. *Answer choices:*

(see index for correct answer)

- a. Right to be forgotten
- b. I-broker
- c. Domain privacy
- d. Intranet

Guidance: level 1

:: Business process ::

A _____ or business method is a collection of related, structured activities or tasks by people or equipment which in a specific sequence produce a service or product for a particular customer or customers. _____ es occur at all organizational levels and may or may not be visible to the customers. A _____ may often be visualized as a flowchart of a sequence of activities with interleaving decision points or as a process matrix of a sequence of activities with relevance rules based on data in the process. The benefits of using _____ es include improved customer satisfaction and improved agility for reacting to rapid market change. Process-oriented organizations break down the barriers of structural departments and try to avoid functional silos.

Exam Probability: **High**

11. *Answer choices:*

(see index for correct answer)

- a. Business process outsourcing
- b. Dynamic business process management
- c. Business operations
- d. Direct store delivery

Guidance: level 1

:: Survey methodology ::

An _____ is a conversation where questions are asked and answers are given. In common parlance, the word "_____" refers to a one-on-one conversation between an _____ er and an _____ ee. The _____ er asks questions to which the _____ ee responds, usually so information may be transferred from _____ ee to _____ er . Sometimes, information can be transferred in both directions. It is a communication, unlike a speech, which produces a one-way flow of information.

Exam Probability: **Low**

12. *Answer choices:*

(see index for correct answer)

- a. Survey research
- b. Interview
- c. American Association for Public Opinion Research
- d. Sampling

Guidance: level 1

:: Marketing by medium ::

_____ or viral advertising is a business strategy that uses existing social networks to promote a product. Its name refers to how consumers spread information about a product with other people in their social networks, much in the same way that a virus spreads from one person to another. It can be delivered by word of mouth or enhanced by the network effects of the Internet and mobile networks.

Exam Probability: **High**

13. *Answer choices:*

(see index for correct answer)

- a. Viral marketing
- b. New media marketing
- c. Direct Text Marketing
- d. Brand infiltration

Guidance: level 1

:: ::

_____ is a brand name associated with the development of the _____ web browser. It is now owned by Verizon Media, a subsidiary of Verizon. The brand belonged to the _____ Communications Corporation, an independent American computer services company, whose headquarters were in Mountain View, California, and later Dulles, Virginia. The browser was once dominant but lost to Internet Explorer and other competitors after the so-called first browser war, its market share falling from more than 90 percent in the mid-1990s to less than 1 percent in 2006.

Exam Probability: **High**

14. *Answer choices:*

(see index for correct answer)

- a. levels of analysis
- b. empathy
- c. cultural
- d. Netscape

Guidance: level 1

:: Data management ::

" _____ " is a field that treats ways to analyze, systematically extract information from, or otherwise deal with data sets that are too large or complex to be dealt with by traditional data-processing application software. Data with many cases offer greater statistical power, while data with higher complexity may lead to a higher false discovery rate. _____ challenges include capturing data, data storage, data analysis, search, sharing, transfer, visualization, querying, updating, information privacy and data source. _____ was originally associated with three key concepts: volume, variety, and velocity. Other concepts later attributed with _____ are veracity and value.

Exam Probability: **High**

15. *Answer choices:*

(see index for correct answer)

- a. Big data
- b. SciDB
- c. Metadata controller
- d. Content management

Guidance: level 1

:: Ergonomics ::

_____ is the design of products, devices, services, or environments for people with disabilities. The concept of accessible design and practice of accessible development ensures both "direct access" and "indirect access" meaning compatibility with a person's assistive technology.

Exam Probability: **High**

16. *Answer choices:*
(see index for correct answer)

- a. Accessibility
- b. Ergonomics in Canada
- c. High-velocity human factors
- d. Ergonomic glove

Guidance: level 1

:: Strategic management ::

_____ is a management term for an element that is necessary for an organization or project to achieve its mission. Alternative terms are key result area and key success factor.

Exam Probability: **High**

17. *Answer choices:*

(see index for correct answer)

- a. customer lock-in
- b. Strategic Technology Plan
- c. strategy implementation
- d. Critical success factor

Guidance: level 1

:: ::

A _____ is a knowledge base website on which users collaboratively modify content and structure directly from the web browser. In a typical _____, text is written using a simplified markup language and often edited with the help of a rich-text editor.

Exam Probability: **Medium**

18. *Answer choices:*

(see index for correct answer)

- a. Wiki
- b. interpersonal communication
- c. open system
- d. surface-level diversity

Guidance: level 1

:: Enterprise architecture ::

Enterprise software, also known as _____ software, is computer software used to satisfy the needs of an organization rather than individual users. Such organizations include businesses, schools, interest-based user groups, clubs, charities, and governments. Enterprise software is an integral part of a information system.

Exam Probability: **High**

19. *Answer choices:*

(see index for correct answer)

- a. Enterprise information security architecture
- b. Enterprise application
- c. Systems architect
- d. Enterprise Collaboration Architecture

Guidance: level 1

:: Data management ::

In business, _____ is a method used to define and manage the critical data of an organization to provide, with data integration, a single point of reference. The data that is mastered may include reference data- the set of permissible values, and the analytical data that supports decision making.

Exam Probability: **Medium**

20. *Answer choices:*

(see index for correct answer)

- a. Master data management
- b. Single customer view
- c. Enterprise data management
- d. Data governance

Guidance: level 1

:: Computer data ::

In computer science, _____ is the ability to access an arbitrary element of a sequence in equal time or any datum from a population of addressable elements roughly as easily and efficiently as any other, no matter how many elements may be in the set. It is typically contrasted to sequential access.

Exam Probability: **Low**

21. *Answer choices:*

(see index for correct answer)

- a. DataPortability
- b. Attribute
- c. Random access
- d. Stream

Guidance: level 1

:: ::

_____ is the function of specifying access rights/privileges to resources, which is related to information security and computer security in general and to access control in particular. More formally, "to authorize" is to define an access policy. For example, human resources staff are normally authorized to access employee records and this policy is usually formalized as access control rules in a computer system. During operation, the system uses the access control rules to decide whether access requests from consumers shall be approved or disapproved. Resources include individual files or an item's data, computer programs, computer devices and functionality provided by computer applications. Examples of consumers are computer users, computer Software and other Hardware on the computer.

Exam Probability: **Medium**

22. *Answer choices:*

(see index for correct answer)

- a. Sarbanes-Oxley act of 2002
- b. cultural
- c. hierarchical perspective
- d. Authorization

Guidance: level 1

:: Identity management ::

_____ is the ability of an individual or group to seclude themselves, or information about themselves, and thereby express themselves selectively. The boundaries and content of what is considered private differ among cultures and individuals, but share common themes. When something is private to a person, it usually means that something is inherently special or sensitive to them. The domain of _____ partially overlaps with security, which can include the concepts of appropriate use, as well as protection of information. _____ may also take the form of bodily integrity.

Exam Probability: **Medium**

23. *Answer choices:*
(see index for correct answer)

- a. Certification on demand
- b. Privacy
- c. Identity 3.0

- d. Directory information tree

Guidance: level 1

:: E-commerce ::

Customer to customer markets provide an innovative way to allow customers to interact with each other. Traditional markets require business to customer relationships, in which a customer goes to the business in order to purchase a product or service. In customer to customer markets, the business facilitates an environment where customers can sell goods or services to each other. Other types of markets include business to business and business to customer.

Exam Probability: **Medium**

24. *Answer choices:*

(see index for correct answer)

- a. Online savings account
- b. Silent commerce
- c. Consumer-to-consumer
- d. E-tendering

Guidance: level 1

:: Computer file formats ::

_____ is a communication protocol for peer-to-peer file sharing which is used to distribute data and electronic files over the Internet.

Exam Probability: **Low**

25. *Answer choices:*

(see index for correct answer)

- a. Be-Music Source
- b. Dynamic-link library
- c. 8SVX
- d. Comtrade

Guidance: level 1

:: Infographics ::

A _____ is a symbolic representation of information according to visualization technique. _____ s have been used since ancient times, but became more prevalent during the Enlightenment. Sometimes, the technique uses a three-dimensional visualization which is then projected onto a two-dimensional surface. The word graph is sometimes used as a synonym for _____ .

Exam Probability: **High**

26. *Answer choices:*

(see index for correct answer)

- a. Diagram
- b. U.S. Route shield
- c. Statistical graphics
- d. Technical drawing

Guidance: level 1

:: Credit cards ::

> A _____ is a payment card issued to users to enable the cardholder to pay a merchant for goods and services based on the cardholder's promise to the card issuer to pay them for the amounts plus the other agreed charges. The card issuer creates a revolving account and grants a line of credit to the cardholder, from which the cardholder can borrow money for payment to a merchant or as a cash advance.

Exam Probability: **Low**

27. *Answer choices:*

(see index for correct answer)

- a. Offshore credit card
- b. Credit card
- c. Wirecard
- d. Payments as a service

Guidance: level 1

:: Payment systems ::

A _____ is any system used to settle financial transactions through the transfer of monetary value. This includes the institutions, instruments, people, rules, procedures, standards, and technologies that make it exchange possible. A common type of _____ is called an operational network that links bank accounts and provides for monetary exchange using bank deposits. Some _____ s also include credit mechanisms, which are essentially a different aspect of payment.

Exam Probability: **Medium**

28. *Answer choices:*

(see index for correct answer)

- a. LevelUp
- b. Bankgiro
- c. Payment system
- d. Sleekpay

Guidance: level 1

:: Marketing ::

_____ , in marketing, manufacturing, call centres and management, is the use of flexible computer-aided manufacturing systems to produce custom output. Such systems combine the low unit costs of mass production processes with the flexibility of individual customization.

Exam Probability: **Medium**

29. *Answer choices:*

(see index for correct answer)

- a. Discounts and allowances
- b. Pharmaceutical marketing
- c. Democratized transactional giving
- d. Buyer decision process

Guidance: level 1

:: Data management ::

_____ is a set of processes and technologies that supports the collection, managing, and publishing of information in any form or medium. When stored and accessed via computers, this information may be more specifically referred to as digital content, or simply as content.

Exam Probability: **Medium**

30. *Answer choices:*

(see index for correct answer)

- a. Archive site
- b. Hierarchical classifier
- c. Content management
- d. Atomicity

Guidance: level 1

:: Transaction processing ::

Transaction processing is information processing in computer science that is divided into individual, indivisible operations called transactions. Each transaction must succeed or fail as a complete unit; it can never be only partially complete.

Exam Probability: **Medium**

31. *Answer choices:*

(see index for correct answer)

- a. Same-day affirmation
- b. Kernel Transaction Manager
- c. Blind write
- d. Consistency model

Guidance: level 1

:: Data analysis ::

_____ , also referred to as text data mining, roughly equivalent to text analytics, is the process of deriving high-quality information from text. High-quality information is typically derived through the devising of patterns and trends through means such as statistical pattern learning. _____ usually involves the process of structuring the input text , deriving patterns within the structured data, and finally evaluation and interpretation of the output. `High quality` in _____ usually refers to some combination of relevance, novelty, and interest. Typical _____ tasks include text categorization, text clustering, concept/entity extraction, production of granular taxonomies, sentiment analysis, document summarization, and entity relation modeling .

Exam Probability: **Medium**

32. *Answer choices:*

(see index for correct answer)

- a. LISREL
- b. PhUSE
- c. Text mining
- d. Variance-stabilizing transformation

Guidance: level 1

:: Computer access control protocols ::

An _____ is a type of computer communications protocol or cryptographic protocol specifically designed for transfer of authentication data between two entities. It allows the receiving entity to authenticate the connecting entity as well as authenticate itself to the connecting entity by declaring the type of information needed for authentication as well as syntax. It is the most important layer of protection needed for secure communication within computer networks.

Exam Probability: **Low**

33. *Answer choices:*

(see index for correct answer)

- a. Kerberos
- b. Simple Authentication and Security Layer
- c. Ticket Granting Ticket
- d. NTLMSSP

Guidance: level 1

:: Geographic information systems ::

_____ is the computational process of transforming a physical address description to a location on the Earth's surface. Reverse _____, on the other hand, converts geographic coordinates to a description of a location, usually the name of a place or an addressable location. _____ relies on a computer representation of address points, the street / road network, together with postal and administrative boundaries.

Exam Probability: **Medium**

34. *Answer choices:*

(see index for correct answer)

- a. Geocoding
- b. Geo-content
- c. Historical geographic information system
- d. South African Spatial Data Infrastructure

Guidance: level 1

:: ::

A _____ is a control panel usually located directly ahead of a vehicle's driver, displaying instrumentation and controls for the vehicle's operation.

Exam Probability: **Low**

35. *Answer choices:*

(see index for correct answer)

- a. functional perspective
- b. empathy
- c. Dashboard
- d. similarity-attraction theory

Guidance: level 1

:: Content management systems ::

_____ is the textual, visual, or aural content that is encountered as part of the user experience on websites. It may include—among other things—text, images, sounds, videos, and animations.

Exam Probability: **Medium**

36. *Answer choices:*

(see index for correct answer)

- a. ConcourseConnect
- b. Web content
- c. Armstrong
- d. Pixie

Guidance: level 1

:: Business process ::

_____ is a discipline in operations management in which people use various methods to discover, model, analyze, measure, improve, optimize, and automate business processes. BPM focuses on improving corporate performance by managing business processes. Any combination of methods used to manage a company's business processes is BPM. Processes can be structured and repeatable or unstructured and variable. Though not required, enabling technologies are often used with BPM.

Exam Probability: **Low**

37. *Answer choices:*

(see index for correct answer)

- a. Business process management
- b. International business development
- c. Business Process Definition Metamodel
- d. Extended Enterprise Modeling Language

Guidance: level 1

:: Network theory ::

A _____ is a social structure made up of a set of social actors, sets of dyadic ties, and other social interactions between actors. The _____ perspective provides a set of methods for analyzing the structure of whole social entities as well as a variety of theories explaining the patterns observed in these structures. The study of these structures uses _____ analysis to identify local and global patterns, locate influential entities, and examine network dynamics.

Exam Probability: **Medium**

38. *Answer choices:*

(see index for correct answer)

- a. Surprise
- b. Widest path problem
- c. Social network
- d. Link analysis

Guidance: level 1

:: Game artificial intelligence ::

In computer science, _____ , sometimes called machine intelligence, is intelligence demonstrated by machines, in contrast to the natural intelligence displayed by humans and animals. Colloquially, the term " _____ " is used to describe machines that mimic "cognitive" functions that humans associate with other human minds, such as "learning" and "problem solving".

Exam Probability: **Medium**

39. *Answer choices:*

(see index for correct answer)

- a. Quiescence search
- b. Proof-number search

- c. Zobrist hashing
- d. Transposition table

Guidance: level 1

:: Data management ::

_____ represents the business objects that contain the most valuable, agreed upon information shared across an organization. It can cover relatively static reference data, transactional, unstructured, analytical, hierarchical and metadata. It is the primary focus of the information technology discipline of _____ management.

Exam Probability: **High**

40. *Answer choices:*

(see index for correct answer)

- a. Scriptella
- b. Australian National Data Service
- c. Data frame
- d. Master data

Guidance: level 1

:: Information systems ::

In artificial intelligence, an _____ is a computer system that emulates the decision-making ability of a human expert. _____ s are designed to solve complex problems by reasoning through bodies of knowledge, represented mainly as if–then rules rather than through conventional procedural code. The first _____ s were created in the 1970s and then proliferated in the 1980s. _____ s were among the first truly successful forms of artificial intelligence software. However, some experts point out that _____ s were not part of true artificial intelligence since they lack the ability to learn autonomously from external data. An _____ is divided into two subsystems: the inference engine and the knowledge base. The knowledge base represents facts and rules. The inference engine applies the rules to the known facts to deduce new facts. Inference engines can also include explanation and debugging abilities.

Exam Probability: **High**

41. *Answer choices:*

(see index for correct answer)

- a. Semantic broker
- b. Intelligent decision support system
- c. DAD-IS
- d. Expert system

Guidance: level 1

:: Infographics ::

A _____ is a graphical representation of data, in which "the data is represented by symbols, such as bars in a bar _____, lines in a line _____, or slices in a pie _____". A _____ can represent tabular numeric data, functions or some kinds of qualitative structure and provides different info.

Exam Probability: **Medium**

42. *Answer choices:*

(see index for correct answer)

- a. No symbol
- b. Nautical chart
- c. Pioneer plaque
- d. Chart

Guidance: level 1

:: Web security exploits ::

A _____ is a baked or cooked food that is small, flat and sweet. It usually contains flour, sugar and some type of oil or fat. It may include other ingredients such as raisins, oats, chocolate chips, nuts, etc.

Exam Probability: **High**

43. *Answer choices:*

(see index for correct answer)

- a. HTTP Strict Transport Security
- b. Cross-site request forgery
- c. Cross-site tracing
- d. Cross-site cooking

Guidance: level 1

:: ::

_____, Inc. is an American online social media and social networking service company based in Menlo Park, California. It was founded by Mark Zuckerberg, along with fellow Harvard College students and roommates Eduardo Saverin, Andrew McCollum, Dustin Moskovitz and Chris Hughes. It is considered one of the Big Four technology companies along with Amazon, Apple, and Google.

Exam Probability: **High**

44. *Answer choices:*
(see index for correct answer)

- a. cultural
- b. Sarbanes-Oxley act of 2002
- c. Facebook
- d. Character

Guidance: level 1

:: Data modeling languages ::

> An entity–relationship model describes interrelated things of interest in a specific domain of knowledge. A basic ER model is composed of entity types and specifies relationships that can exist between entities.

Exam Probability: **High**

45. *Answer choices:*

(see index for correct answer)

- a. Interface description language
- b. Data Base Task Group
- c. Information Object Class
- d. Entity-relationship

Guidance: level 1

:: ::

A _____ is a computer network that interconnects computers within a limited area such as a residence, school, laboratory, university campus or office building. By contrast, a wide area network not only covers a larger geographic distance, but also generally involves leased telecommunication circuits.

Exam Probability: **Medium**

46. *Answer choices:*

(see index for correct answer)

- a. interpersonal communication
- b. Local Area Network
- c. corporate values
- d. surface-level diversity

Guidance: level 1

:: Information technology management ::

_____ within quality management systems and information technology systems is a process—either formal or informal—used to ensure that changes to a product or system are introduced in a controlled and coordinated manner. It reduces the possibility that unnecessary changes will be introduced to a system without forethought, introducing faults into the system or undoing changes made by other users of software. The goals of a _____ procedure usually include minimal disruption to services, reduction in back-out activities, and cost-effective utilization of resources involved in implementing change.

Exam Probability: **High**

47. *Answer choices:*

(see index for correct answer)

- a. Lean IT
- b. Digital asset management
- c. Change control
- d. Belarc

Guidance: level 1

:: Data management ::

_____ s or data _____ s are computer languages used to make queries in databases and information systems.

Exam Probability: **Medium**

48. *Answer choices:*

(see index for correct answer)

- a. Long-lived transaction
- b. Grid-oriented storage
- c. NewSQL
- d. Inverted index

Guidance: level 1

:: ::

_____ Holdings, Inc. is an American company operating a worldwide online payments system that supports online money transfers and serves as an electronic alternative to traditional paper methods like checks and money orders. The company operates as a payment processor for online vendors, auction sites, and many other commercial users, for which it charges a fee in exchange for benefits such as one-click transactions and password memory. _____'s payment system, also called _____, is considered a type of payment rail.

Exam Probability: **High**

49. *Answer choices:*

(see index for correct answer)

- a. Character
- b. information systems assessment
- c. surface-level diversity
- d. PayPal

Guidance: level 1

:: Google services ::

_____ is a time-management and scheduling calendar service developed by Google. It became available in beta release April 13, 2006, and in general release in July 2009, on the web and as mobile apps for the Android and iOS platforms.

Exam Probability: **High**

50. *Answer choices:*

(see index for correct answer)

- a. Google Alerts
- b. Google Moderator
- c. Google Calendar
- d. Google Map Maker

Guidance: level 1

:: Information science ::

In discourse-based grammatical theory, _____ is any tracking of referential information by speakers. Information may be new, just introduced into the conversation; given, already active in the speakers' consciousness; or old, no longer active. The various types of activation, and how these are defined, are model-dependent.

Exam Probability: **High**

51. *Answer choices:*

(see index for correct answer)

- a. Secondary source
- b. Informative modelling
- c. Interviewer effect
- d. Information flow

Guidance: level 1

:: Information technology management ::

_____ is a collective term for all approaches to prepare, support and help individuals, teams, and organizations in making organizational change. The most common change drivers include: technological evolution, process reviews, crisis, and consumer habit changes; pressure from new business entrants, acquisitions, mergers, and organizational restructuring. It includes methods that redirect or redefine the use of resources, business process, budget allocations, or other modes of operation that significantly change a company or organization. Organizational _____ considers the full organization and what needs to change, while _____ may be used solely to refer to how people and teams are affected by such organizational transition. It deals with many different disciplines, from behavioral and social sciences to information technology and business solutions.

Exam Probability: **Medium**

52. *Answer choices:*

(see index for correct answer)

- a. Web operations
- b. DocSTAR
- c. Mobile document access
- d. IQuate

Guidance: level 1

:: ::

Sustainability is the process of people maintaining change in a balanced environment, in which the exploitation of resources, the direction of investments, the orientation of technological development and institutional change are all in harmony and enhance both current and future potential to meet human needs and aspirations. For many in the field, sustainability is defined through the following interconnected domains or pillars: environment, economic and social, which according to Fritjof Capra is based on the principles of Systems Thinking. Sub-domains of _____ development have been considered also: cultural, technological and political. While _____ development may be the organizing principle for sustainability for some, for others, the two terms are paradoxical . _____ development is the development that meets the needs of the present without compromising the ability of future generations to meet their own needs. Brundtland Report for the World Commission on Environment and Development introduced the term of _____ development.

Exam Probability: **High**

53. *Answer choices:*
(see index for correct answer)

- a. Character

- b. hierarchical perspective
- c. Sustainable
- d. imperative

Guidance: level 1

:: Data management ::

Given organizations' increasing dependency on information technology to run their operations, Business continuity planning covers the entire organization, and Disaster recovery focuses on IT.

Exam Probability: **High**

54. *Answer choices:*

(see index for correct answer)

- a. Disaster recovery plan
- b. Data integration
- c. Data independence
- d. Linked data

Guidance: level 1

:: Statistical laws ::

In statistics and business, a _____ of some distributions of numbers is the portion of the distribution having a large number of occurrences far from the "head" or central part of the distribution. The distribution could involve popularities, random numbers of occurrences of events with various probabilities, etc. The term is often used loosely, with no definition or arbitrary definition, but precise definitions are possible.

Exam Probability: **Medium**

55. *Answer choices:*
(see index for correct answer)

- a. Law of truly large numbers
- b. Pareto principle
- c. Long tail
- d. Rank-size distribution

Guidance: level 1

:: Data transmission ::

In telecommunication a _____ is the means of connecting one location to another for the purpose of transmitting and receiving digital information. It can also refer to a set of electronics assemblies, consisting of a transmitter and a receiver and the interconnecting data telecommunication circuit. These are governed by a link protocol enabling digital data to be transferred from a data source to a data sink.

Exam Probability: **High**

56. *Answer choices:*

(see index for correct answer)

- a. Transport Sample Protocol
- b. Message format
- c. Data link
- d. MISMO

Guidance: level 1

:: Data management ::

_____ is "data [information] that provides information about other data". Many distinct types of _____ exist, among these descriptive _____, structural _____, administrative _____, reference _____ and statistical _____.

Exam Probability: **Medium**

57. *Answer choices:*

(see index for correct answer)

- a. DAMA
- b. Metadata
- c. Data set

- d. Novell Storage Manager

Guidance: level 1

:: IT risk management ::

_____ involves a set of policies, tools and procedures to enable the recovery or continuation of vital technology infrastructure and systems following a natural or human-induced disaster. _____ focuses on the IT or technology systems supporting critical business functions, as opposed to business continuity, which involves keeping all essential aspects of a business functioning despite significant disruptive events. _____ can therefore be considered as a subset of business continuity.

Exam Probability: **Medium**

58. *Answer choices:*

(see index for correct answer)

- a. Incident response team
- b. Information assurance
- c. Disaster recovery

Guidance: level 1

:: Information retrieval ::

_____ is the practice of making content from multiple enterprise-type sources, such as databases and intranets, searchable to a defined audience.

Exam Probability: **Medium**

59. *Answer choices:*

(see index for correct answer)

- a. Datanet
- b. Enterprise search
- c. Binary Independence Model
- d. Natural language user interface

Guidance: level 1

Marketing

Marketing is the study and management of exchange relationships. Marketing is the business process of creating relationships with and satisfying customers. With its focus on the customer, marketing is one of the premier components of business management.

Marketing is defined by the American Marketing Association as "the activity, set of institutions, and processes for creating, communicating, delivering, and exchanging offerings that have value for customers, clients, partners, and society at large."

:: Advertising by type ::

_____ or advertising war is an advertisement in which a particular product, or service, specifically mentions a competitor by name for the express purpose of showing why the competitor is inferior to the product naming it. Also referred to as "knocking copy", it is loosely defined as advertising where "the advertised brand is explicitly compared with one or more competing brands and the comparison is obvious to the audience."

Exam Probability: **High**

1. *Answer choices:*

(see index for correct answer)

- a. Space advertising
- b. Clover campaign
- c. Contextual advertising
- d. Comparative advertising

Guidance: level 1

:: Business law ::

A _____ is an arrangement where parties, known as partners, agree to cooperate to advance their mutual interests. The partners in a _____ may be individuals, businesses, interest-based organizations, schools, governments or combinations. Organizations may partner to increase the likelihood of each achieving their mission and to amplify their reach. A _____ may result in issuing and holding equity or may be only governed by a contract.

Exam Probability: **Medium**

2. *Answer choices:*

(see index for correct answer)

- a. Contract failure
- b. Ordinary resolution
- c. Holder
- d. Registered agent

Guidance: level 1

:: Brand management ::

In marketing, _____ is the analysis and planning on how a brand is perceived in the market. Developing a good relationship with the target market is essential for _____ . Tangible elements of _____ include the product itself; its look, price, and packaging, etc. The intangible elements are the experiences that the consumers share with the brand, and also the relationships they have with the brand. A brand manager would oversee all aspects of the consumer's brand association as well as relationships with members of the supply chain.

Exam Probability: **High**

3. *Answer choices:*

(see index for correct answer)

- a. Brand architecture
- b. Brand preference
- c. VCU Brandcenter
- d. Brand management

Guidance: level 1

:: Television terminology ::

A _____ organization, also known as a non-business entity, not-for-profit organization, or _____ institution, is dedicated to furthering a particular social cause or advocating for a shared point of view. In economic terms, it is an organization that uses its surplus of the revenues to further achieve its ultimate objective, rather than distributing its income to the organization's shareholders, leaders, or members. _____ s are tax exempt or charitable, meaning they do not pay income tax on the money that they receive for their organization. They can operate in religious, scientific, research, or educational settings.

Exam Probability: **Low**

4. *Answer choices:*

(see index for correct answer)

- a. Satellite television
- b. multiplexing
- c. Nonprofit
- d. not-for-profit

Guidance: level 1

:: ::

In the broadest sense, _____ is any practice which contributes to the sale of products to a retail consumer. At a retail in-store level, _____ refers to the variety of products available for sale and the display of those products in such a way that it stimulates interest and entices customers to make a purchase.

Exam Probability: **Medium**

5. *Answer choices:*

(see index for correct answer)

- a. Merchandising
- b. similarity-attraction theory
- c. personal values
- d. co-culture

Guidance: level 1

:: ::

In business and engineering, new _____ covers the complete process of bringing a new product to market. A central aspect of NPD is product design, along with various business considerations. New _____ is described broadly as the transformation of a market opportunity into a product available for sale. The product can be tangible or intangible, though sometimes services and other processes are distinguished from "products." NPD requires an understanding of customer needs and wants, the competitive environment, and the nature of the market. Cost, time and quality are the main variables that drive customer needs. Aiming at these three variables, innovative companies develop continuous practices and strategies to better satisfy customer requirements and to increase their own market share by a regular development of new products. There are many uncertainties and challenges which companies must face throughout the process. The use of best practices and the elimination of barriers to communication are the main concerns for the management of the NPD.

Exam Probability: **Low**

6. *Answer choices:*

(see index for correct answer)

- a. similarity-attraction theory
- b. Product development
- c. functional perspective
- d. open system

Guidance: level 1

:: Contract law ::

A _____ is a legally-binding agreement which recognises and governs the rights and duties of the parties to the agreement. A _____ is legally enforceable because it meets the requirements and approval of the law. An agreement typically involves the exchange of goods, services, money, or promises of any of those. In the event of breach of _____ , the law awards the injured party access to legal remedies such as damages and cancellation.

Exam Probability: **Medium**

7. *Answer choices:*

(see index for correct answer)

- a. Prenuptial agreement
- b. Illegal agreement
- c. Extended warranty
- d. Contract

Guidance: level 1

:: Commodities ::

In economics, a _____ is an economic good or service that has full or substantial fungibility: that is, the market treats instances of the good as equivalent or nearly so with no regard to who produced them. Most commodities are raw materials, basic resources, agricultural, or mining products, such as iron ore, sugar, or grains like rice and wheat. Commodities can also be mass-produced unspecialized products such as chemicals and computer memory.

Exam Probability: **Low**

8. *Answer choices:*

(see index for correct answer)

- a. IRely
- b. Commodity pathway diversion
- c. Commodity
- d. Commoditization

Guidance: level 1

:: Advertising ::

_____ is the behavioral and cognitive process of selectively concentrating on a discrete aspect of information, whether deemed subjective or objective, while ignoring other perceivable information. It is a state of arousal. It is the taking possession by the mind in clear and vivid form of one out of what seem several simultaneous objects or trains of thought. Focalization, the concentration of consciousness, is of its essence. _____ has also been described as the allocation of limited cognitive processing resources.

Exam Probability: **Low**

9. *Answer choices:*

(see index for correct answer)

- a. Media buying
- b. Sex in advertising
- c. Targeted advertising
- d. Attention

Guidance: level 1

:: Cultural appropriation ::

> _____ is a social and economic order that encourages the acquisition of goods and services in ever-increasing amounts. With the industrial revolution, but particularly in the 20th century, mass production led to an economic crisis: there was overproduction—the supply of goods would grow beyond consumer demand, and so manufacturers turned to planned obsolescence and advertising to manipulate consumer spending. In 1899, a book on _____ published by Thorstein Veblen, called The Theory of the Leisure Class, examined the widespread values and economic institutions emerging along with the widespread "leisure time" in the beginning of the 20th century. In it Veblen "views the activities and spending habits of this leisure class in terms of conspicuous and vicarious consumption and waste. Both are related to the display of status and not to functionality or usefulness."

Exam Probability: **High**

10. *Answer choices:*

(see index for correct answer)

- a. Neotribalism
- b. California Indian Song

- c. Consumerism
- d. Washington Redskins Original Americans Foundation

Guidance: level 1

:: Marketing analytics ::

_____ is a long-term, forward-looking approach to planning with the fundamental goal of achieving a sustainable competitive advantage. Strategic planning involves an analysis of the company's strategic initial situation prior to the formulation, evaluation and selection of market-oriented competitive position that contributes to the company's goals and marketing objectives.

Exam Probability: **High**

11. *Answer choices:*

(see index for correct answer)

- a. Gross rating point
- b. Mission-driven marketing
- c. Sumall
- d. Marketing strategy

Guidance: level 1

:: Commerce ::

_____ relates to "the exchange of goods and services, especially on a large scale". It includes legal, economic, political, social, cultural and technological systems that operate in a country or in international trade.

Exam Probability: **High**

12. *Answer choices:*

(see index for correct answer)

- a. Deal transaction
- b. Emerging Markets Index
- c. European Retail Round Table
- d. Commerce

Guidance: level 1

:: E-commerce ::

_____ is the activity of buying or selling of products on online services or over the Internet. Electronic commerce draws on technologies such as mobile commerce, electronic funds transfer, supply chain management, Internet marketing, online transaction processing, electronic data interchange, inventory management systems, and automated data collection systems.

Exam Probability: **High**

13. *Answer choices:*

(see index for correct answer)

- a. Online flower delivery
- b. E-commerce
- c. Acquirer
- d. Online wallet

Guidance: level 1

:: ::

_____ , also referred to as orthostasis, is a human position in which the body is held in an upright position and supported only by the feet.

Exam Probability: **High**

14. *Answer choices:*

(see index for correct answer)

- a. similarity-attraction theory
- b. interpersonal communication
- c. hierarchical perspective
- d. Standing

Guidance: level 1

:: Marketing ::

_____ is the percentage of a market accounted for by a specific entity. In a survey of nearly 200 senior marketing managers, 67% responded that they found the revenue- "dollar _____" metric very useful, while 61% found "unit _____" very useful.

Exam Probability: **High**

15. *Answer choices:*

(see index for correct answer)

- a. Market share
- b. Drug coupon
- c. Point of difference
- d. Beat-sheet

Guidance: level 1

:: Stock market ::

The _____ of a corporation is all of the shares into which ownership of the corporation is divided. In American English, the shares are commonly known as "_____s". A single share of the _____ represents fractional ownership of the corporation in proportion to the total number of shares. This typically entitles the _____ holder to that fraction of the company's earnings, proceeds from liquidation of assets, or voting power, often dividing these up in proportion to the amount of money each _____ holder has invested. Not all _____ is necessarily equal, as certain classes of _____ may be issued for example without voting rights, with enhanced voting rights, or with a certain priority to receive profits or liquidation proceeds before or after other classes of shareholders.

Exam Probability: **Medium**

16. *Answer choices:*

(see index for correct answer)

- a. GXG Markets
- b. Shadow stock
- c. Leading stock
- d. Immediate or cancel

Guidance: level 1

:: Pricing ::

_____ is a pricing strategy in which the selling price is determined by adding a specific amount markup to a product's unit cost. An alternative pricing method is value-based pricing.

Exam Probability: **High**

17. *Answer choices:*

(see index for correct answer)

- a. Net metering
- b. Dynamic pricing
- c. Water tariff
- d. Cost-plus pricing

Guidance: level 1

:: Data collection ::

A _____ is an utterance which typically functions as a request for information. _____ s can thus be understood as a kind of illocutionary act in the field of pragmatics or as special kinds of propositions in frameworks of formal semantics such as alternative semantics or inquisitive semantics. The information requested is expected to be provided in the form of an answer. _____ s are often conflated with interrogatives, which are the grammatical forms typically used to achieve them. Rhetorical _____ s, for example, are interrogative in form but may not be considered true _____ s as they are not expected to be answered. Conversely, non-interrogative grammatical structures may be considered _____ s as in the case of the imperative sentence "tell me your name".

Exam Probability: **High**

18. *Answer choices:*

(see index for correct answer)

- a. Natural experiment
- b. Question
- c. Provenance
- d. IPUMS

Guidance: level 1

:: ::

_____ is a term frequently used in marketing. It is a measure of how products and services supplied by a company meet or surpass customer expectation. _____ is defined as "the number of customers, or percentage of total customers, whose reported experience with a firm, its products, or its services exceeds specified satisfaction goals."

Exam Probability: **High**

19. *Answer choices:*

(see index for correct answer)

- a. personal values
- b. functional perspective
- c. levels of analysis
- d. Customer satisfaction

Guidance: level 1

:: ::

In sales, commerce and economics, a _____ is the recipient of a good, service, product or an idea - obtained from a seller, vendor, or supplier via a financial transaction or exchange for money or some other valuable consideration.

Exam Probability: **High**

20. *Answer choices:*

(see index for correct answer)

- a. hierarchical
- b. Customer
- c. information systems assessment
- d. imperative

Guidance: level 1

:: Production economics ::

In microeconomics, _____ are the cost advantages that enterprises obtain due to their scale of operation, with cost per unit of output decreasing with increasing scale.

Exam Probability: **Low**

21. *Answer choices:*

(see index for correct answer)

- a. Isoquant
- b. Economies of scale
- c. Marginal product
- d. Marginal cost of capital schedule

Guidance: level 1

:: Public relations ::

_____ is the public visibility or awareness for any product, service or company. It may also refer to the movement of information from its source to the general public, often but not always via the media. The subjects of _____ include people, goods and services, organizations, and works of art or entertainment.

Exam Probability: **Medium**

22. *Answer choices:*

(see index for correct answer)

- a. Defense Video %26 Imagery Distribution System
- b. Zakazukha
- c. Publicity
- d. Public Relations Inquiry

Guidance: level 1

:: ::

An _____ is an area of the production, distribution, or trade, and consumption of goods and services by different agents. Understood in its broadest sense, `The _____ is defined as a social domain that emphasize the practices, discourses, and material expressions associated with the production, use, and management of resources`. Economic agents can be individuals, businesses, organizations, or governments. Economic transactions occur when two parties agree to the value or price of the transacted good or service, commonly expressed in a certain currency. However, monetary transactions only account for a small part of the economic domain.

Exam Probability: **Medium**

23. *Answer choices:*

(see index for correct answer)

- a. Economy
- b. personal values
- c. process perspective

- d. levels of analysis

Guidance: level 1

:: ::

A _____ is a professional who provides expert advice in a particular area such as security, management, education, accountancy, law, human resources, marketing, finance, engineering, science or any of many other specialized fields.

Exam Probability: **High**

24. *Answer choices:*
(see index for correct answer)

- a. cultural
- b. deep-level diversity
- c. open system
- d. Consultant

Guidance: level 1

:: ::

Management is the administration of an organization, whether it is a business, a not-for-profit organization, or government body. Management includes the activities of setting the strategy of an organization and coordinating the efforts of its employees to accomplish its objectives through the application of available resources, such as financial, natural, technological, and human resources. The term "management" may also refer to those people who manage an organization.

Exam Probability: **Low**

25. *Answer choices:*

(see index for correct answer)

- a. personal values
- b. similarity-attraction theory
- c. levels of analysis
- d. hierarchical

Guidance: level 1

:: Product management ::

_____ or brand stretching is a marketing strategy in which a firm marketing a product with a well-developed image uses the same brand name in a different product category. The new product is called a spin-off. Organizations use this strategy to increase and leverage brand equity. An example of a _____ is Jello-gelatin creating Jello pudding pops. It increases awareness of the brand name and increases profitability from offerings in more than one product category.

Exam Probability: **High**

26. *Answer choices:*

(see index for correct answer)

- a. Electronic registration mark
- b. Brand equity
- c. Obsolescence
- d. Brand extension

Guidance: level 1

:: Library science ::

_____ refers to data which is collected by someone who is someone other than the user. Common sources of _____ for social science include censuses, information collected by government departments, organizational records and data that was originally collected for other research purposes. Primary data, by contrast, are collected by the investigator conducting the research.

Exam Probability: **High**

27. *Answer choices:*

(see index for correct answer)

- a. Annotated bibliography
- b. Sourcebook

- c. NEFLIN
- d. University of Chicago Graduate Library School

Guidance: level 1

:: Marketing ::

A _____ is an overall experience of a customer that distinguishes an organization or product from its rivals in the eyes of the customer. _____ s are used in business, marketing, and advertising. Name _____ s are sometimes distinguished from generic or store _____ s.

Exam Probability: **High**

28. *Answer choices:*
(see index for correct answer)

- a. DirectIndustry
- b. Mandatory labelling
- c. Azerbaijan Marketing Society
- d. Gimmick

Guidance: level 1

:: Survey methodology ::

A _____ is the procedure of systematically acquiring and recording information about the members of a given population. The term is used mostly in connection with national population and housing _____ es; other common _____ es include agriculture, business, and traffic _____ es. The United Nations defines the essential features of population and housing _____ es as "individual enumeration, universality within a defined territory, simultaneity and defined periodicity", and recommends that population _____ es be taken at least every 10 years. United Nations recommendations also cover _____ topics to be collected, official definitions, classifications and other useful information to co-ordinate international practice.

Exam Probability: **Low**

29. *Answer choices:*
(see index for correct answer)

- a. American Association for Public Opinion Research
- b. World Association for Public Opinion Research
- c. Group concept mapping
- d. Census

Guidance: level 1

:: Marketing ::

_____ uses different marketing channels and tools in combination: Marketing communication channels focus on any way a business communicates a message to its desired market, or the market in general. A marketing communication tool can be anything from: advertising, personal selling, direct marketing, sponsorship, communication, and promotion to public relations.

Exam Probability: **High**

30. *Answer choices:*

(see index for correct answer)

- a. Buy one, get one free
- b. Nutraceutical
- c. Digital brand engagement
- d. Green marketing

Guidance: level 1

:: Consumer behaviour ::

_____ refers to the ability of a company or product to retain its customers over some specified period. High _____ means customers of the product or business tend to return to, continue to buy or in some other way not defect to another product or business, or to non-use entirely. Selling organizations generally attempt to reduce customer defections. _____ starts with the first contact an organization has with a customer and continues throughout the entire lifetime of a relationship and successful retention efforts take this entire lifecycle into account. A company's ability to attract and retain new customers is related not only to its product or services, but also to the way it services its existing customers, the value the customers actually generate as a result of utilizing the solutions, and the reputation it creates within and across the marketplace.

Exam Probability: **High**

31. *Answer choices:*

(see index for correct answer)

- a. Blissful ignorance effect
- b. Denomination effect
- c. Daniel Starch
- d. Cocooning

Guidance: level 1

:: Monopoly (economics) ::

A _____ is a form of intellectual property that gives its owner the legal right to exclude others from making, using, selling, and importing an invention for a limited period of years, in exchange for publishing an enabling public disclosure of the invention. In most countries _____ rights fall under civil law and the _____ holder needs to sue someone infringing the _____ in order to enforce his or her rights. In some industries _____ s are an essential form of competitive advantage; in others they are irrelevant.

Exam Probability: **Low**

32. *Answer choices:*

(see index for correct answer)

- a. De facto monopoly
- b. Patent portfolio
- c. Patent
- d. National Competition Policy

Guidance: level 1

:: Marketing techniques ::

The _____ or unique selling point is a marketing concept first
proposed as a theory to explain a pattern in successful advertising campaigns
of the early 1940s. The USP states that such campaigns made unique propositions
to customers that convinced them to switch brands. The term was developed by
television advertising pioneer Rosser Reeves of Ted Bates & Company.
Theodore Levitt, a professor at Harvard Business School, suggested that,
"Differentiation is one of the most important strategic and tactical activities
in which companies must constantly engage." The term has been used to describe
one`s "personal brand" in the marketplace. Today, the term is used in other
fields or just casually to refer to any aspect of an object that differentiates
it from similar objects.

Exam Probability: **Medium**

33. *Answer choices:*

(see index for correct answer)

- a. Relevant space
- b. Unique selling proposition
- c. Not sold in stores
- d. Flyposting

Guidance: level 1

:: Asset ::

In financial accounting, an _____ is any resource owned by the business. Anything tangible or intangible that can be owned or controlled to produce value and that is held by a company to produce positive economic value is an _____. Simply stated, _____s represent value of ownership that can be converted into cash. The balance sheet of a firm records the monetary value of the _____s owned by that firm. It covers money and other valuables belonging to an individual or to a business.

Exam Probability: **Medium**

34. *Answer choices:*

(see index for correct answer)

- a. Current asset
- b. Fixed asset

Guidance: level 1

:: Evaluation methods ::

_____ is a scientific method of observation to gather non-numerical data. This type of research "refers to the meanings, concepts definitions, characteristics, metaphors, symbols, and description of things" and not to their "counts or measures." This research answers why and how a certain phenomenon may occur rather than how often. _____ approaches are employed across many academic disciplines, focusing particularly on the human elements of the social and natural sciences; in less academic contexts, areas of application include qualitative market research, business, service demonstrations by non-profits, and journalism.

Exam Probability: **Medium**

35. *Answer choices:*

(see index for correct answer)

- a. Video ethnography
- b. Qualitative research
- c. Quantitative research
- d. Most significant change technique

Guidance: level 1

:: ::

_____ is the process of gathering and measuring information on targeted variables in an established system, which then enables one to answer relevant questions and evaluate outcomes. _____ is a component of research in all fields of study including physical and social sciences, humanities, and business. While methods vary by discipline, the emphasis on ensuring accurate and honest collection remains the same. The goal for all _____ is to capture quality evidence that allows analysis to lead to the formulation of convincing and credible answers to the questions that have been posed.

Exam Probability: **Low**

36. *Answer choices:*

(see index for correct answer)

- a. levels of analysis
- b. personal values
- c. Sarbanes-Oxley act of 2002
- d. Data collection

Guidance: level 1

:: ::

Competition arises whenever at least two parties strive for a goal which cannot be shared: where one's gain is the other's loss.

Exam Probability: **Medium**

37. *Answer choices:*

(see index for correct answer)

- a. open system
- b. Sarbanes-Oxley act of 2002
- c. Competitor
- d. process perspective

Guidance: level 1

:: ::

The _____ is a U.S. business-focused, English-language international daily newspaper based in New York City. The Journal, along with its Asian and European editions, is published six days a week by Dow Jones & Company, a division of News Corp. The newspaper is published in the broadsheet format and online. The Journal has been printed continuously since its inception on July 8, 1889, by Charles Dow, Edward Jones, and Charles Bergstresser.

Exam Probability: **Medium**

38. *Answer choices:*

(see index for correct answer)

- a. Wall Street Journal
- b. deep-level diversity
- c. empathy
- d. similarity-attraction theory

Guidance: level 1

:: Consumer theory ::

_____ is the quantity of a good that consumers are willing and able to purchase at various prices during a given period of time.

Exam Probability: **Medium**

39. *Answer choices:*

(see index for correct answer)

- a. intertemporal substitution
- b. Autonomous consumption
- c. Joint demand
- d. Consumer sovereignty

Guidance: level 1

:: Workplace ::

_____ is asystematic determination of a subject's merit, worth and significance, using criteria governed by a set of standards. It can assist an organization, program, design, project or any other intervention or initiative to assess any aim, realisable concept/proposal, or any alternative, to help in decision-making; or to ascertain the degree of achievement or value in regard to the aim and objectives and results of any such action that has been completed. The primary purpose of _____, in addition to gaining insight into prior or existing initiatives, is to enable reflection and assist in the identification of future change.

Exam Probability: **High**

40. *Answer choices:*
(see index for correct answer)

- a. Discrimination based on hair texture
- b. Open allocation
- c. Workplace harassment

- d. Occupational stress

Guidance: level 1

:: Internet privacy ::

An _____ is a private network accessible only to an organization's staff. Often, a wide range of information and services are available on an organization's internal _____ that are unavailable to the public, unlike the Internet. A company-wide _____ can constitute an important focal point of internal communication and collaboration, and provide a single starting point to access internal and external resources. In its simplest form, an _____ is established with the technologies for local area networks and wide area networks . Many modern _____ s have search engines, user profiles, blogs, mobile apps with notifications, and events planning within their infrastructure.

Exam Probability: **High**

41. *Answer choices:*

(see index for correct answer)

- a. Cypherpunk
- b. Anonymity application
- c. Intranet
- d. Digitalcourage

Guidance: level 1

:: ::

A _____ is a discussion or informational website published on the World Wide Web consisting of discrete, often informal diary-style text entries. Posts are typically displayed in reverse chronological order, so that the most recent post appears first, at the top of the web page. Until 2009, _____ s were usually the work of a single individual, occasionally of a small group, and often covered a single subject or topic. In the 2010s, "multi-author _____ s" emerged, featuring the writing of multiple authors and sometimes professionally edited. MABs from newspapers, other media outlets, universities, think tanks, advocacy groups, and similar institutions account for an increasing quantity of _____ traffic. The rise of Twitter and other "micro _____ ging" systems helps integrate MABs and single-author _____ s into the news media. _____ can also be used as a verb, meaning to maintain or add content to a _____ .

Exam Probability: **Low**

42. *Answer choices:*

(see index for correct answer)

- a. functional perspective
- b. personal values
- c. similarity-attraction theory
- d. interpersonal communication

Guidance: level 1

:: Marketing ::

_____ is "commercial competition characterized by the repeated cutting of prices below those of competitors". One competitor will lower its price, then others will lower their prices to match. If one of them reduces their price again, a new round of reductions starts. In the short term, _____s are good for buyers, who can take advantage of lower prices. Often they are not good for the companies involved because the lower prices reduce profit margins and can threaten their survival.

Exam Probability: **Low**

43. *Answer choices:*

(see index for correct answer)

- a. Price war
- b. Adobe Marketing Cloud
- c. Gold party
- d. Decoy effect

Guidance: level 1

:: ::

_____ is the collection of techniques, skills, methods, and processes used in the production of goods or services or in the accomplishment of objectives, such as scientific investigation. _____ can be the knowledge of techniques, processes, and the like, or it can be embedded in machines to allow for operation without detailed knowledge of their workings. Systems applying _____ by taking an input, changing it according to the system's use, and then producing an outcome are referred to as _____ systems or technological systems.

Exam Probability: **Medium**

44. *Answer choices:*

(see index for correct answer)

- a. corporate values
- b. similarity-attraction theory
- c. Technology
- d. co-culture

Guidance: level 1

:: Marketing terminology ::

_____ is used in marketing to describe the inability to assess the value gained from engaging in an activity using any tangible evidence. It is often used to describe services where there is no tangible product that the customer can purchase, that can be seen or touched.

Exam Probability: **Low**

45. *Answer choices:*

(see index for correct answer)

- a. Aspirational age
- b. Intangibility
- c. Name program
- d. Oscar bait

Guidance: level 1

:: Business models ::

_____ es are privately owned corporations, partnerships, or sole proprietorships that have fewer employees and/or less annual revenue than a regular-sized business or corporation. Businesses are defined as "small" in terms of being able to apply for government support and qualify for preferential tax policy varies depending on the country and industry. _____ es range from fifteen employees under the Australian Fair Work Act 2009, fifty employees according to the definition used by the European Union, and fewer than five hundred employees to qualify for many U.S. _____ Administration programs. While _____ es can also be classified according to other methods, such as annual revenues, shipments, sales, assets, or by annual gross or net revenue or net profits, the number of employees is one of the most widely used measures.

Exam Probability: **Low**

46. *Answer choices:*

(see index for correct answer)

- a. InnovationXchange
- b. Collective business system
- c. Small business
- d. Brainsworking

Guidance: level 1

:: Data management ::

_____ is a form of intellectual property that grants the creator of an original creative work an exclusive legal right to determine whether and under what conditions this original work may be copied and used by others, usually for a limited term of years. The exclusive rights are not absolute but limited by limitations and exceptions to _____ law, including fair use. A major limitation on _____ on ideas is that _____ protects only the original expression of ideas, and not the underlying ideas themselves.

Exam Probability: **Low**

47. *Answer choices:*

(see index for correct answer)

- a. SQL/PSM
- b. Data governance
- c. Operational database

- d. Cleansing and Conforming Data

Guidance: level 1

:: Management ::

A _____ describes the rationale of how an organization creates, delivers, and captures value, in economic, social, cultural or other contexts. The process of _____ construction and modification is also called _____ innovation and forms a part of business strategy.

Exam Probability: **Medium**

48. *Answer choices:*
(see index for correct answer)

- a. Business model
- b. Planning
- c. Task-oriented and relationship-oriented leadership
- d. Gemba

Guidance: level 1

:: Strategic alliances ::

A _____ is an agreement between two or more parties to pursue a set of agreed upon objectives needed while remaining independent organizations. A _____ will usually fall short of a legal partnership entity, agency, or corporate affiliate relationship. Typically, two companies form a _____ when each possesses one or more business assets or have expertise that will help the other by enhancing their businesses. _____s can develop in outsourcing relationships where the parties desire to achieve long-term win-win benefits and innovation based on mutually desired outcomes.

Exam Probability: **Medium**

49. *Answer choices:*

(see index for correct answer)

- a. Management contract
- b. International joint venture
- c. Strategic alliance
- d. Bridge Alliance

Guidance: level 1

:: Direct marketing ::

_____ is a method of direct marketing in which a salesperson solicits prospective customers to buy products or services, either over the phone or through a subsequent face to face or Web conferencing appointment scheduled during the call. _____ can also include recorded sales pitches programmed to be played over the phone via automatic dialing.

Exam Probability: **High**

50. *Answer choices:*

(see index for correct answer)

- a. Telemarketing
- b. Flyer
- c. Time Reading Program
- d. Mailing list

Guidance: level 1

:: Decision theory ::

A _____ is a deliberate system of principles to guide decisions and achieve rational outcomes. A _____ is a statement of intent, and is implemented as a procedure or protocol. Policies are generally adopted by a governance body within an organization. Policies can assist in both subjective and objective decision making. Policies to assist in subjective decision making usually assist senior management with decisions that must be based on the relative merits of a number of factors, and as a result are often hard to test objectively, e.g. work-life balance _____. In contrast policies to assist in objective decision making are usually operational in nature and can be objectively tested, e.g. password _____.

Exam Probability: **Low**

51. *Answer choices:*

(see index for correct answer)

- a. Emotional bias
- b. Decision field theory
- c. Consensus decision-making
- d. Policy

Guidance: level 1

_____ is both a research area and a practical skill encompassing the ability of an individual or organization to "lead" or guide other individuals, teams, or entire organizations. Specialist literature debates various viewpoints, contrasting Eastern and Western approaches to _____, and also United States versus European approaches. U.S. academic environments define _____ as "a process of social influence in which a person can enlist the aid and support of others in the accomplishment of a common task".

Exam Probability: **High**

52. *Answer choices:*

(see index for correct answer)

- a. Leadership
- b. surface-level diversity
- c. information systems assessment
- d. corporate values

Guidance: level 1

:: Market research ::

_____ is an organized effort to gather information about target markets or customers. It is a very important component of business strategy. The term is commonly interchanged with marketing research; however, expert practitioners may wish to draw a distinction, in that marketing research is concerned specifically about marketing processes, while _____ is concerned specifically with markets.

Exam Probability: **High**

53. *Answer choices:*

(see index for correct answer)

- a. Media Technology Monitor
- b. Market research
- c. ISO 20252
- d. Monroe Mendelsohn Research

Guidance: level 1

:: Marketing ::

_____, sometimes called trigger-based or event-driven marketing, is a marketing strategy that uses two-way communication channels to allow consumers to connect with a company directly. Although this exchange can take place in person, in the last decade it has increasingly taken place almost exclusively online through email, social media, and blogs.

Exam Probability: **Low**

54. *Answer choices:*

(see index for correct answer)

- a. Product planning
- b. Boston matrix
- c. Market intelligence
- d. Interactive marketing

Guidance: level 1

:: ::

A _____ is an organization, usually a group of people or a company, authorized to act as a single entity and recognized as such in law. Early incorporated entities were established by charter. Most jurisdictions now allow the creation of new _____ s through registration.

Exam Probability: **High**

55. *Answer choices:*

(see index for correct answer)

- a. similarity-attraction theory
- b. levels of analysis
- c. corporate values
- d. interpersonal communication

Guidance: level 1

:: Market research ::

_____ is "the process or set of processes that links the producers, customers, and end users to the marketer through information used to identify and define marketing opportunities and problems; generate, refine, and evaluate marketing actions; monitor marketing performance; and improve understanding of marketing as a process. _____ specifies the information required to address these issues, designs the method for collecting information, manages and implements the data collection process, analyzes the results, and communicates the findings and their implications."

Exam Probability: **High**

56. *Answer choices:*

(see index for correct answer)

- a. Mall-intercept personal interview
- b. IDDEA

- c. Marketing research
- d. Industry analyst

Guidance: level 1

:: ::

_____ is change in the heritable characteristics of biological populations over successive generations. These characteristics are the expressions of genes that are passed on from parent to offspring during reproduction. Different characteristics tend to exist within any given population as a result of mutation, genetic recombination and other sources of genetic variation. _____ occurs when _____ ary processes such as natural selection and genetic drift act on this variation, resulting in certain characteristics becoming more common or rare within a population. It is this process of _____ that has given rise to biodiversity at every level of biological organisation, including the levels of species, individual organisms and molecules.

Exam Probability: **Medium**

57. *Answer choices:*

(see index for correct answer)

- a. functional perspective
- b. co-culture
- c. Character
- d. surface-level diversity

Guidance: level 1

:: Evaluation methods ::

In natural and social sciences, and sometimes in other fields, _____ is the systematic empirical investigation of observable phenomena via statistical, mathematical, or computational techniques. The objective of _____ is to develop and employ mathematical models, theories, and hypotheses pertaining to phenomena. The process of measurement is central to _____ because it provides the fundamental connection between empirical observation and mathematical expression of quantitative relationships.

Exam Probability: **High**

58. *Answer choices:*
(see index for correct answer)

- a. Poll average
- b. Event correlation
- c. Alternative assessment
- d. Quantitative research

Guidance: level 1

:: Production and manufacturing ::

_____ consists of organization-wide efforts to "install and make permanent climate where employees continuously improve their ability to provide on demand products and services that customers will find of particular value." "Total" emphasizes that departments in addition to production are obligated to improve their operations; "management" emphasizes that executives are obligated to actively manage quality through funding, training, staffing, and goal setting. While there is no widely agreed-upon approach, TQM efforts typically draw heavily on the previously developed tools and techniques of quality control. TQM enjoyed widespread attention during the late 1980s and early 1990s before being overshadowed by ISO 9000, Lean manufacturing, and Six Sigma.

Exam Probability: **High**

59. *Answer choices:*

(see index for correct answer)

- a. Copacker
- b. Total Quality Management
- c. EFQM Excellence Model
- d. Enterprise control

Guidance: level 1

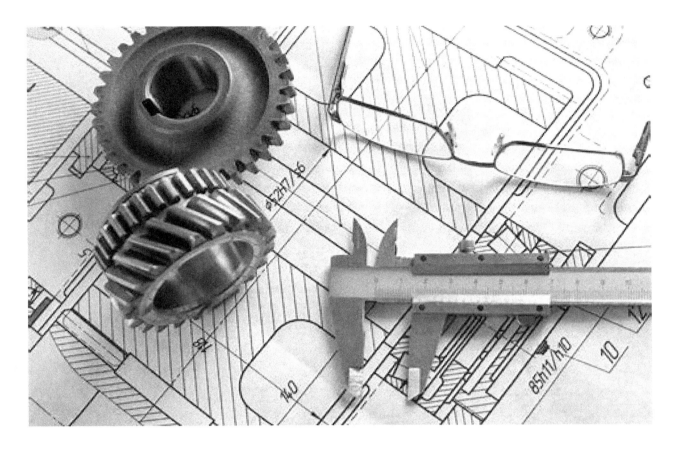

Manufacturing

Manufacturing is the production of merchandise for use or sale using labor and machines, tools, chemical and biological processing, or formulation. The term may refer to a range of human activity, from handicraft to high tech, but is most commonly applied to industrial design , in which raw materials are transformed into finished goods on a large scale. Such finished goods may be sold to other manufacturers for the production of other, more complex products, such as aircraft, household appliances, furniture, sports equipment or automobiles, or sold to wholesalers, who in turn sell them to retailers, who then sell them to end users and consumers.

:: Materials science ::

An _____ is a polymer with viscoelasticity and very weak intermolecular forces, and generally low Young's modulus and high failure strain compared with other materials. The term, a portmanteau of elastic polymer, is often used interchangeably with rubber, although the latter is preferred when referring to vulcanisates. Each of the monomers which link to form the polymer is usually a compound of several elements among carbon, hydrogen, oxygen and silicon. _____s are amorphous polymers maintained above their glass transition temperature, so that considerable molecular reconformation, without breaking of covalent bonds, is feasible. At ambient temperatures, such rubbers are thus relatively soft and deformable. Their primary uses are for seals, adhesives and molded flexible parts. Application areas for different types of rubber are manifold and cover segments as diverse as tires, soles for shoes, and damping and insulating elements. The importance of these rubbers can be judged from the fact that global revenues are forecast to rise to US$56 billion in 2020.

Exam Probability: **Low**

1. *Answer choices:*

(see index for correct answer)

- a. Negative thermal expansion
- b. Bi-isotropic material
- c. Fracture toughening mechanisms
- d. Eugene Podkletnov

Guidance: level 1

:: ::

Catalysis is the process of increasing the rate of a chemical reaction by adding a substance known as a _____ , which is not consumed in the catalyzed reaction and can continue to act repeatedly. Because of this, only very small amounts of _____ are required to alter the reaction rate in principle.

Exam Probability: **Medium**

2. *Answer choices:*

(see index for correct answer)

- a. interpersonal communication
- b. functional perspective
- c. surface-level diversity
- d. Catalyst

Guidance: level 1

:: Distribution, retailing, and wholesaling ::

The _____ is a distribution channel phenomenon in which forecasts yield supply chain inefficiencies. It refers to increasing swings in inventory in response to shifts in customer demand as one moves further up the supply chain. The concept first appeared in Jay Forrester's Industrial Dynamics and thus it is also known as the Forrester effect. The _____ was named for the way the amplitude of a whip increases down its length. The further from the originating signal, the greater the distortion of the wave pattern. In a similar manner, forecast accuracy decreases as one moves upstream along the supply chain. For example, many consumer goods have fairly consistent consumption at retail but this signal becomes more chaotic and unpredictable as the focus moves away from consumer purchasing behavior.

Exam Probability: **Low**

3. *Answer choices:*

(see index for correct answer)

- a. Demand modeling
- b. Pallet racking
- c. Bullwhip effect
- d. Cost to serve

Guidance: level 1

:: Production and manufacturing ::

_____ is a concept in purchasing and project management for securing the quality and timely delivery of goods and components.

Exam Probability: **High**

4. *Answer choices:*

(see index for correct answer)

- a. Factory Instrumentation Protocol
- b. Division of labour
- c. Process layout
- d. SynqNet

Guidance: level 1

:: Water ::

_____ is a transparent, tasteless, odorless, and nearly colorless chemical substance, which is the main constituent of Earth's streams, lakes, and oceans, and the fluids of most living organisms. It is vital for all known forms of life, even though it provides no calories or organic nutrients. Its chemical formula is H2O, meaning that each of its molecules contains one oxygen and two hydrogen atoms, connected by covalent bonds. _____ is the name of the liquid state of H2O at standard ambient temperature and pressure. It forms precipitation in the form of rain and aerosols in the form of fog. Clouds are formed from suspended droplets of _____ and ice, its solid state. When finely divided, crystalline ice may precipitate in the form of snow. The gaseous state of _____ is steam or _____ vapor. _____ moves continually through the _____ cycle of evaporation, transpiration, condensation, precipitation, and runoff, usually reaching the sea.

Exam Probability: **High**

5. *Answer choices:*

(see index for correct answer)

- a. Water
- b. Water efficiency
- c. Capillary fringe
- d. Orange County Water District

Guidance: level 1

:: Costs ::

In process improvement efforts, _____ or cost of quality is a means to quantify the total cost of quality-related efforts and deficiencies. It was first described by Armand V. Feigenbaum in a 1956 Harvard Business Review article.

Exam Probability: **High**

6. *Answer choices:*

(see index for correct answer)

- a. Flyaway cost
- b. Customer Cost
- c. Khozraschyot
- d. Quality costs

Guidance: level 1

:: Promotion and marketing communications ::

The _____ of American Manufacturers, now ThomasNet, is an online platform for supplier discovery and product sourcing in the US and Canada. It was once known as the "big green books" and "Thomas Registry", and was a multi-volume directory of industrial product information covering 650,000 distributors, manufacturers and service companies within 67,000-plus industrial categories that is now published on ThomasNet.

Exam Probability: **Low**

7. *Answer choices:*

(see index for correct answer)

- a. media kit
- b. Thomas Register
- c. Shop fitting
- d. News propaganda

Guidance: level 1

:: Asset ::

In financial accounting, an _____ is any resource owned by the business. Anything tangible or intangible that can be owned or controlled to produce value and that is held by a company to produce positive economic value is an _____ . Simply stated, _____ s represent value of ownership that can be converted into cash . The balance sheet of a firm records the monetary value of the _____ s owned by that firm. It covers money and other valuables belonging to an individual or to a business.

Exam Probability: **High**

8. *Answer choices:*

(see index for correct answer)

- a. Fixed asset
- b. Current asset

Guidance: level 1

:: ::

A _____ or till is a mechanical or electronic device for registering and calculating transactions at a point of sale. It is usually attached to a drawer for storing cash and other valuables. A modern _____ is usually attached to a printer that can print out receipts for record-keeping purposes.

Exam Probability: **High**

9. *Answer choices:*

(see index for correct answer)

- a. deep-level diversity
- b. cultural
- c. process perspective
- d. Cash register

Guidance: level 1

:: Decision theory ::

_____ is a method developed in Japan beginning in 1966 to help transform the voice of the customer into engineering characteristics for a product. Yoji Akao, the original developer, described QFD as a "method to transform qualitative user demands into quantitative parameters, to deploy the functions forming quality, and to deploy methods for achieving the design quality into subsystems and component parts, and ultimately to specific elements of the manufacturing process." The author combined his work in quality assurance and quality control points with function deployment used in value engineering.

Exam Probability: **Medium**

10. *Answer choices:*

(see index for correct answer)

- a. Shared decision-making
- b. Quality function deployment
- c. Quantum cognition

- d. Polychotomous key

Guidance: level 1

:: Direct marketing ::

_____ Inc. is an American privately owned multi-level marketing company. According to Direct Selling News, _____ was the sixth largest network marketing company in the world in 2018, with a wholesale volume of US$3.25 billion. _____ is based in Addison, Texas, outside Dallas. The company was founded by _____ Ash in 1963. Richard Rogers, _____'s son, is the chairman, and David Holl is president and was named CEO in 2006.

Exam Probability: **Medium**

11. *Answer choices:*

(see index for correct answer)

- a. Mailing list
- b. Mary Kay
- c. Time Reading Program
- d. Direct Marketing Association

Guidance: level 1

:: ::

Some scenarios associate "this kind of planning" with learning "life skills".Schedules are necessary, or at least useful, in situations where individuals need to know what time they must be at a specific location to receive a specific service, and where people need to accomplish a set of goals within a set time period.

Exam Probability: **Medium**

12. *Answer choices:*

(see index for correct answer)

- a. surface-level diversity
- b. deep-level diversity
- c. empathy
- d. hierarchical

Guidance: level 1

:: Product development ::

In business and engineering, _____ covers the complete process of bringing a new product to market. A central aspect of NPD is product design, along with various business considerations. _____ is described broadly as the transformation of a market opportunity into a product available for sale. The product can be tangible or intangible , though sometimes services and other processes are distinguished from "products." NPD requires an understanding of customer needs and wants, the competitive environment, and the nature of the market.Cost, time and quality are the main variables that drive customer needs. Aiming at these three variables, innovative companies develop continuous practices and strategies to better satisfy customer requirements and to increase their own market share by a regular development of new products. There are many uncertainties and challenges which companies must face throughout the process. The use of best practices and the elimination of barriers to communication are the main concerns for the management of the NPD .

Exam Probability: **High**

13. *Answer choices:*

(see index for correct answer)

- a. Product line extension
- b. Material selection
- c. Minimum viable product
- d. Design for assembly

Guidance: level 1

:: Marketing ::

_____ or stock is the goods and materials that a business holds for the ultimate goal of resale.

Exam Probability: **High**

14. *Answer choices:*

(see index for correct answer)

- a. All-commodity volume
- b. Neuromarketing
- c. Demonstrator model
- d. Buyer decision process

Guidance: level 1

:: Project management ::

Some scenarios associate "this kind of planning" with learning "life skills". _____ s are necessary, or at least useful, in situations where individuals need to know what time they must be at a specific location to receive a specific service, and where people need to accomplish a set of goals within a set time period.

Exam Probability: **High**

15. *Answer choices:*

(see index for correct answer)

- a. Theory Z
- b. A Guide to the Project Management Body of Knowledge
- c. Test and evaluation master plan
- d. Fast-track construction

Guidance: level 1

:: ::

A _____ consists of an orchestrated and repeatable pattern of business activity enabled by the systematic organization of resources into processes that transform materials, provide services, or process information. It can be depicted as a sequence of operations, the work of a person or group, the work of an organization of staff, or one or more simple or complex mechanisms.

Exam Probability: **High**

16. *Answer choices:*
(see index for correct answer)

- a. functional perspective
- b. interpersonal communication
- c. information systems assessment
- d. Workflow

Guidance: level 1

:: Costs ::

_____ is the process used by companies to reduce their costs and increase their profits. Depending on a company's services or product, the strategies can vary. Every decision in the product development process affects cost.

Exam Probability: **Medium**

17. *Answer choices:*

(see index for correct answer)

- a. Total cost
- b. Explicit cost
- c. Cost of products sold
- d. Manufacturing cost

Guidance: level 1

:: Unit operations ::

_____ is the process of separating the components or substances from a liquid mixture by using selective boiling and condensation. _____ may result in essentially complete separation, or it may be a partial separation that increases the concentration of selected components in the mixture. In either case, the process exploits differences in the volatility of the mixture's components. In industrial chemistry, _____ is a unit operation of practically universal importance, but it is a physical separation process, not a chemical reaction.

Exam Probability: **Medium**

18. *Answer choices:*
(see index for correct answer)

- a. Heat transfer
- b. Unit Operations of Chemical Engineering
- c. Clearing factor
- d. Settling

Guidance: level 1

:: Production economics ::

_____ is the creation of a whole that is greater than the simple sum of its parts. The term _____ comes from the Attic Greek word sea synergia from synergos, , meaning "working together".

Exam Probability: **Medium**

19. Answer choices:

(see index for correct answer)

- a. Productivity Alpha
- b. Fragmentation
- c. Capacity utilization
- d. Synergy

Guidance: level 1

:: Inventory ::

The _____ is the level of inventory which triggers an action to replenish that particular inventory stock. It is a minimum amount of an item which a firm holds in stock, such that, when stock falls to this amount, the item must be reordered. It is normally calculated as the forecast usage during the replenishment lead time plus safety stock. In the EOQ model, it was assumed that there is no time lag between ordering and procuring of materials. Therefore the _____ for replenishing the stocks occurs at that level when the inventory level drops to zero and because instant delivery by suppliers, the stock level bounce back.

Exam Probability: **High**

20. Answer choices:

(see index for correct answer)

- a. Spare part
- b. Inventory bounce

- c. Lower of cost or market
- d. Reorder point

Guidance: level 1

:: Management ::

_____ is a term used in business and Information Technology to describe the in-depth process of capturing customer's expectations, preferences and aversions. Specifically, the _____ is a market research technique that produces a detailed set of customer wants and needs, organized into a hierarchical structure, and then prioritized in terms of relative importance and satisfaction with current alternatives. _____ studies typically consist of both qualitative and quantitative research steps. They are generally conducted at the start of any new product, process, or service design initiative in order to better understand the customer's wants and needs, and as the key input for new product definition, Quality Function Deployment, and the setting of detailed design specifications.

Exam Probability: **Low**

21. *Answer choices:*
(see index for correct answer)

- a. Public sector consulting
- b. Omnex
- c. Business workflow analysis
- d. Voice of the customer

Guidance: level 1

:: Process management ::

A _____ is a diagram commonly used in chemical and process engineering to indicate the general flow of plant processes and equipment. The PFD displays the relationship between major equipment of a plant facility and does not show minor details such as piping details and designations. Another commonly used term for a PFD is a flowsheet.

Exam Probability: **High**

22. *Answer choices:*

(see index for correct answer)

- a. Business process modeling
- b. Turnaround
- c. Conformance checking
- d. Integrated business planning

Guidance: level 1

:: Business process ::

A committee is a body of one or more persons that is subordinate to a deliberative assembly. Usually, the assembly sends matters into a committee as a way to explore them more fully than would be possible if the assembly itself were considering them. Committees may have different functions and their type of work differ depending on the type of the organization and its needs.

Exam Probability: **Medium**

23. *Answer choices:*

(see index for correct answer)

- a. Misuse case
- b. Business operations
- c. Software ecosystem
- d. Steering committee

Guidance: level 1

:: Marketing ::

_____ or stock control can be broadly defined as "the activity of checking a shop's stock." However, a more focused definition takes into account the more science-based, methodical practice of not only verifying a business' inventory but also focusing on the many related facets of inventory management "within an organisation to meet the demand placed upon that business economically." Other facets of _____ include supply chain management, production control, financial flexibility, and customer satisfaction. At the root of _____, however, is the _____ problem, which involves determining when to order, how much to order, and the logistics of those decisions.

Exam Probability: **High**

24. *Answer choices:*

(see index for correct answer)

- a. Mandatory labelling
- b. MARC USA
- c. Inventory control
- d. Online ethnography

Guidance: level 1

:: Management ::

_____ is a category of business activity made possible by software tools that aim to provide customers with both independence from vendors and better means for engaging with vendors. These same tools can also apply to individuals' relations with other institutions and organizations.

Exam Probability: **Medium**

25. *Answer choices:*

(see index for correct answer)

- a. Double linking
- b. Vendor relationship management
- c. Reverse innovation
- d. Mobile sales enablement

Guidance: level 1

:: ::

_____ is the quantity of three-dimensional space enclosed by a closed surface, for example, the space that a substance or shape occupies or contains. _____ is often quantified numerically using the SI derived unit, the cubic metre. The _____ of a container is generally understood to be the capacity of the container; i. e., the amount of fluid that the container could hold, rather than the amount of space the container itself displaces. Three dimensional mathematical shapes are also assigned _____ s. _____ s of some simple shapes, such as regular, straight-edged, and circular shapes can be easily calculated using arithmetic formulas. _____ s of complicated shapes can be calculated with integral calculus if a formula exists for the shape's boundary. One-dimensional figures and two-dimensional shapes are assigned zero _____ in the three-dimensional space.

Exam Probability: **Low**

26. *Answer choices:*

(see index for correct answer)

- a. personal values
- b. surface-level diversity
- c. Character
- d. co-culture

Guidance: level 1

:: Business planning ::

_____ is a critical component to the successful delivery of any project, programme or activity. A stakeholder is any individual, group or organization that can affect, be affected by, or perceive itself to be affected by a programme.

Exam Probability: **Low**

27. *Answer choices:*

(see index for correct answer)

- a. operational planning
- b. Community Futures
- c. Stakeholder management
- d. Open Options Corporation

Guidance: level 1

:: Management ::

_____ , also known as natural process limits, are horizontal lines drawn on a statistical process control chart, usually at a distance of ±3 standard deviations of the plotted statistic from the statistic's mean.

Exam Probability: **High**

28. *Answer choices:*

(see index for correct answer)

- a. Social business model
- b. Control limits
- c. Operations research
- d. Crisis plan

Guidance: level 1

:: Quality ::

The _____ , formerly the _____ Control, is a knowledge-based global community of quality professionals, with nearly 80,000 members dedicated to promoting and advancing quality tools, principles, and practices in their workplaces and communities.

Exam Probability: **High**

29. *Answer choices:*

(see index for correct answer)

- a. Root cause analysis
- b. Dualistic Petri nets
- c. American Society for Quality
- d. Software Engineering Process Group

Guidance: level 1

:: Production and manufacturing ::

_____ is a comprehensive and rigorous industrial process by which a previously sold, leased, used, worn or non-functional product or part is returned to a 'like-new' or 'better-than-new' condition, from both a quality and performance perspective, through a controlled, reproducible and sustainable process.

Exam Probability: **High**

30. *Answer choices:*

(see index for correct answer)

- a. Corrective and preventive action
- b. Economic dispatch
- c. Computer-aided process planning
- d. Remanufacturing

Guidance: level 1

:: Management ::

_____ is the identification, evaluation, and prioritization of risks followed by coordinated and economical application of resources to minimize, monitor, and control the probability or impact of unfortunate events or to maximize the realization of opportunities.

Exam Probability: **Medium**

31. *Answer choices:*

(see index for correct answer)

- a. Communications management
- b. Gemba
- c. Decentralized decision-making
- d. PhD in management

Guidance: level 1

:: Industrial organization ::

In economics, specifically general equilibrium theory, a perfect market is defined by several idealizing conditions, collectively called _____ . In theoretical models where conditions of _____ hold, it has been theoretically demonstrated that a market will reach an equilibrium in which the quantity supplied for every product or service, including labor, equals the quantity demanded at the current price. This equilibrium would be a Pareto optimum.

Exam Probability: **Medium**

32. *Answer choices:*

(see index for correct answer)

- a. Path dependence
- b. Quaternary sector of the economy
- c. American system of manufacturing
- d. Worldwide Responsible Accredited Production

Guidance: level 1

:: Business process ::

_____ is the value to an enterprise which is derived from the techniques, procedures, and programs that implement and enhance the delivery of goods and services. _____ is one of the three components of structural capital, itself a component of intellectual capital. _____ can be seen as the value of processes to any entity, whether for profit or not-for profit, but is most commonly used in reference to for-profit entities.

Exam Probability: **High**

33. *Answer choices:*

(see index for correct answer)

- a. Business process outsourcing
- b. Closure by stealth
- c. Real-time enterprise
- d. ProcessEdge

Guidance: level 1

:: Manufacturing ::

_____ or lean production, often simply "lean", is a systematic method for the minimization of waste within a manufacturing system without sacrificing productivity, which can cause problems. Lean also takes into account waste created through overburden and waste created through unevenness in work loads . Working from the perspective of the client who consumes a product or service, "value" is any action or process that a customer would be willing to pay for.

Exam Probability: **Medium**

34. *Answer choices:*

(see index for correct answer)

- a. IEC 61511
- b. Enterprise appliance transaction module
- c. Lean manufacturing
- d. Build to stock

Guidance: level 1

:: Sampling (statistics) ::

_____ uses statistical sampling to determine whether to accept or reject a production lot of material. It has been a common quality control technique used in industry. It is usually done as products leaves the factory, or in some cases even within the factory. Most often a producer supplies a consumer a number of items and a decision to accept or reject the items is made by determining the number of defective items in a sample from the lot. The lot is accepted if the number of defects falls below where the acceptance number or otherwise the lot is rejected.

Exam Probability: **Low**

35. *Answer choices:*

(see index for correct answer)

- a. Time use survey
- b. Acceptance sampling
- c. Stratified sampling
- d. Empirical evidence

Guidance: level 1

:: Chemical processes ::

_____ is the understanding and application of the fundamental principles and laws of nature that allow us to transform raw material and energy into products that are useful to society, at an industrial level. By taking advantage of the driving forces of nature such as pressure, temperature and concentration gradients, as well as the law of conservation of mass, process engineers can develop methods to synthesize and purify large quantities of desired chemical products. _____ focuses on the design, operation, control, optimization and intensification of chemical, physical, and biological processes. _____ encompasses a vast range of industries, such as agriculture, automotive, biotechnical, chemical, food, material development, mining, nuclear, petrochemical, pharmaceutical, and software development. The application of systematic computer-based methods to _____ is "process systems engineering".

Exam Probability: **Medium**

36. *Answer choices:*

(see index for correct answer)

- a. Starve-fed
- b. Ion plating

- c. Process engineering
- d. Nitrophosphate process

Guidance: level 1

:: Occupational safety and health ::

_____ is a chemical element with symbol Pb and atomic number 82. It is a heavy metal that is denser than most common materials. _____ is soft and malleable, and also has a relatively low melting point. When freshly cut, _____ is silvery with a hint of blue; it tarnishes to a dull gray color when exposed to air. _____ has the highest atomic number of any stable element and three of its isotopes are endpoints of major nuclear decay chains of heavier elements.

Exam Probability: **Low**

37. *Answer choices:*
(see index for correct answer)

- a. Lead
- b. Chlorine
- c. 1,2-Dibromo-3-chloropropane
- d. Global road safety for workers

Guidance: level 1

:: Production and manufacturing ::

_____ was a management-led program to eliminate defects in industrial production that enjoyed brief popularity in American industry from 1964 to the early 1970s. Quality expert Philip Crosby later incorporated it into his "Absolutes of Quality Management" and it enjoyed a renaissance in the American automobile industry—as a performance goal more than as a program—in the 1990s. Although applicable to any type of enterprise, it has been primarily adopted within supply chains wherever large volumes of components are being purchased .

Exam Probability: **Medium**

38. *Answer choices:*

(see index for correct answer)

- a. Zero Defects
- b. STEP-NC
- c. Resource Breakdown
- d. Continuous production

Guidance: level 1

:: Production economics ::

In economics and related disciplines, a _____ is a cost in making any economic trade when participating in a market.

Exam Probability: **High**

39. *Answer choices:*

(see index for correct answer)

- a. Ramp up
- b. Productivity Alpha
- c. Capitalist mode of production
- d. Transaction cost

Guidance: level 1

:: Waste ::

_____ are unwanted or unusable materials. _____ is any substance which is discarded after primary use, or is worthless, defective and of no use. A by-product by contrast is a joint product of relatively minor economic value. A _____ product may become a by-product, joint product or resource through an invention that raises a _____ product's value above zero.

Exam Probability: **Low**

40. *Answer choices:*

(see index for correct answer)

- a. Waste heat
- b. Trash Inc: The Secret Life of Garbage

- c. Sharps waste
- d. Waste

Guidance: level 1

:: Information technology management ::

_____ concerns a cycle of organizational activity: the acquisition of information from one or more sources, the custodianship and the distribution of that information to those who need it, and its ultimate disposition through archiving or deletion.

Exam Probability: **Medium**

41. *Answer choices:*

(see index for correct answer)

- a. ODMA
- b. Device Management Forum
- c. Information management
- d. COBIT

Guidance: level 1

:: Accounting source documents ::

A _____ is a commercial document and first official offer issued by a buyer to a seller indicating types, quantities, and agreed prices for products or services. It is used to control the purchasing of products and services from external suppliers. _____ s can be an essential part of enterprise resource planning system orders.

Exam Probability: **Medium**

42. *Answer choices:*

(see index for correct answer)

- a. Purchase order
- b. Parcel audit
- c. Invoice
- d. Credit memorandum

Guidance: level 1

:: Lean manufacturing ::

_____ is the Sino-Japanese word for "improvement". In business, _____ refers to activities that continuously improve all functions and involve all employees from the CEO to the assembly line workers. It also applies to processes, such as purchasing and logistics, that cross organizational boundaries into the supply chain. It has been applied in healthcare, psychotherapy, life-coaching, government, and banking.

Exam Probability: **Medium**

43. *Answer choices:*

(see index for correct answer)

- a. No value added
- b. 5S
- c. Kaizen
- d. Agent-assisted automation

Guidance: level 1

:: Distribution, retailing, and wholesaling ::

> _____ measures the performance of a system. Certain goals are defined and the _____ gives the percentage to which those goals should be achieved. Fill rate is different from _____ .

Exam Probability: **Medium**

44. *Answer choices:*

(see index for correct answer)

- a. Service level
- b. Filling station
- c. Sales variance
- d. Slab-O-Concrete

Guidance: level 1

:: Management ::

Business _____ is a discipline in operations management in which people use various methods to discover, model, analyze, measure, improve, optimize, and automate business processes. BPM focuses on improving corporate performance by managing business processes. Any combination of methods used to manage a company's business processes is BPM. Processes can be structured and repeatable or unstructured and variable. Though not required, enabling technologies are often used with BPM.

Exam Probability: **Low**

45. *Answer choices:*
(see index for correct answer)

- a. Responsible autonomy
- b. Competitive heterogeneity
- c. Process management
- d. Task-oriented and relationship-oriented leadership

Guidance: level 1

:: Information technology management ::

The term _____ is used to refer to periods when a system is unavailable. _____ or outage duration refers to a period of time that a system fails to provide or perform its primary function. Reliability, availability, recovery, and unavailability are related concepts. The unavailability is the proportion of a time-span that a system is unavailable or offline. This is usually a result of the system failing to function because of an unplanned event, or because of routine maintenance.

Exam Probability: **Low**

46. *Answer choices:*

(see index for correct answer)

- a. Downtime
- b. GESMES/TS
- c. Information model
- d. Mobile business development

Guidance: level 1

:: Management ::

_____ is a process by which entities review the quality of all factors involved in production. ISO 9000 defines _____ as "A part of quality management focused on fulfilling quality requirements".

Exam Probability: **High**

47. *Answer choices:*

(see index for correct answer)

- a. Corporate foresight
- b. Managerial hubris
- c. Process management
- d. Management styles

Guidance: level 1

:: Management ::

In inventory management, _____ is the order quantity that minimizes the total holding costs and ordering costs. It is one of the oldest classical production scheduling models. The model was developed by Ford W. Harris in 1913, but R. H. Wilson, a consultant who applied it extensively, and K. Andler are given credit for their in-depth analysis.

Exam Probability: **High**

48. *Answer choices:*

(see index for correct answer)

- a. Peer pressure
- b. Economic order quantity
- c. Participative decision-making
- d. Sensemaking

Guidance: level 1

:: Management ::

_____ is a formal technique useful where many possible courses of action are competing for attention. In essence, the problem-solver estimates the benefit delivered by each action, then selects a number of the most effective actions that deliver a total benefit reasonably close to the maximal possible one.

Exam Probability: **High**

49. *Answer choices:*

(see index for correct answer)

- a. Project management information system
- b. Semiconductor consolidation
- c. Design management
- d. Value migration

Guidance: level 1

:: Costs ::

The _____ is computed by dividing the total cost of goods available for sale by the total units available for sale. This gives a weighted-average unit cost that is applied to the units in the ending inventory.

Exam Probability: **High**

50. *Answer choices:*

(see index for correct answer)

- a. Average variable cost
- b. Opportunity cost of capital
- c. Prospective costs
- d. Sliding scale fees

Guidance: level 1

:: Management accounting ::

"_____ s are the structural determinants of the cost of an activity, reflecting any linkages or interrelationships that affect it". Therefore we could assume that the _____ s determine the cost behavior within the activities, reflecting the links that these have with other activities and relationships that affect them.

Exam Probability: **Medium**

51. *Answer choices:*

(see index for correct answer)

- a. Cost accounting
- b. Certified Management Accountants of Canada
- c. Operating profit margin
- d. Cost driver

Guidance: level 1

:: Management accounting ::

_____ are costs that are not directly accountable to a cost object . _____ may be either fixed or variable. _____ include administration, personnel and security costs. These are those costs which are not directly related to production. Some _____ may be overhead. But some overhead costs can be directly attributed to a project and are direct costs.

Exam Probability: **Medium**

52. *Answer choices:*

(see index for correct answer)

- a. Certified Management Accountant
- b. Chartered Institute of Management Accountants
- c. Notional profit
- d. Indirect costs

Guidance: level 1

:: Quality ::

A _____ is an initiating cause of either a condition or a causal chain that leads to an outcome or effect of interest. The term denotes the earliest, most basic, `deepest`, cause for a given behavior; most often a fault. The idea is that you can only see an error by its manifest signs. Those signs can be widespread, multitudinous, and convoluted, whereas the _____ leading to them often is a lot simpler.

Exam Probability: **Medium**

53. *Answer choices:*
(see index for correct answer)

- a. Quality of life
- b. Quality by Design
- c. Root cause
- d. Market Driven Quality

Guidance: level 1

:: Project management ::

_____ is a marketing activity that does an aggregate plan for the production process, in advance of 6 to 18 months, to give an idea to management as to what quantity of materials and other resources are to be procured and when, so that the total cost of operations of the organization is kept to the minimum over that period.

Exam Probability: **High**

54. *Answer choices:*

(see index for correct answer)

- a. P3M3
- b. RationalPlan
- c. Aggregate planning
- d. Milestone

Guidance: level 1

:: ::

An _____ is a company that produces parts and equipment that may be marketed by another manufacturer. For example, Foxconn, a Taiwanese electronics contract manufacturing company, which produces a variety of parts and equipment for companies such as Apple Inc., Dell, Google, Huawei, Nintendo, etc., is the largest OEM company in the world by both scale and revenue.

Exam Probability: **Low**

55. *Answer choices:*

(see index for correct answer)

- a. process perspective
- b. Sarbanes-Oxley act of 2002
- c. Original equipment manufacturer
- d. functional perspective

Guidance: level 1

:: Project management ::

A _____ is a source or supply from which a benefit is produced and it has some utility. _____ s can broadly be classified upon their availability—they are classified into renewable and non-renewable _____ s.Examples of non renewable _____ s are coal ,crude oil natural gas nuclear energy etc. Examples of renewable _____ s are air,water,wind,solar energy etc. They can also be classified as actual and potential on the basis of level of development and use, on the basis of origin they can be classified as biotic and abiotic, and on the basis of their distribution, as ubiquitous and localized . An item becomes a _____ with time and developing technology. Typically, _____ s are materials, energy, services, staff, knowledge, or other assets that are transformed to produce benefit and in the process may be consumed or made unavailable. Benefits of _____ utilization may include increased wealth, proper functioning of a system, or enhanced well-being. From a human perspective a natural _____ is anything obtained from the environment to satisfy human needs and wants. From a broader biological or ecological perspective a _____ satisfies the needs of a living organism .

Exam Probability: **High**

56. *Answer choices:*

(see index for correct answer)

- a. Project planning
- b. Resource
- c. Dependency
- d. Small-scale project management

Guidance: level 1

:: Fault-tolerant computer systems ::

_____ decision-making is a group decision-making process in which group members develop, and agree to support a decision in the best interest of the whole group or common goal. _____ may be defined professionally as an acceptable resolution, one that can be supported, even if not the "favourite" of each individual. It has its origin in the Latin word consensus, which is from consentio meaning literally feel together. It is used to describe both the decision and the process of reaching a decision. _____ decision-making is thus concerned with the process of deliberating and finalizing a decision, and the social, economic, legal, environmental and political effects of applying this process.

Exam Probability: **Low**

57. *Answer choices:*

(see index for correct answer)

- a. FlockDB

- b. Virtual synchrony
- c. Consensus
- d. Fencing

Guidance: level 1

:: Non-parametric statistics ::

A _____ is an accurate representation of the distribution of numerical data. It is an estimate of the probability distribution of a continuous variable and was first introduced by Karl Pearson. It differs from a bar graph, in the sense that a bar graph relates two variables, but a _____ relates only one. To construct a _____ , the first step is to "bin" the range of values—that is, divide the entire range of values into a series of intervals—and then count how many values fall into each interval. The bins are usually specified as consecutive, non-overlapping intervals of a variable. The bins must be adjacent, and are often of equal size.

Exam Probability: **High**

58. *Answer choices:*
(see index for correct answer)

- a. Van der Waerden test
- b. Sign test
- c. Semiparametric regression
- d. Friedman test

Guidance: level 1

:: Help desk ::

Data center management is the collection of tasks performed by those responsible for managing ongoing operation of a data center This includes Business service management and planning for the future.

Exam Probability: **Medium**

59. *Answer choices:*

(see index for correct answer)

- a. SysAid Technologies
- b. Technical support
- c. Vitalyst
- d. AetherPal

Guidance: level 1

Commerce

Commerce relates to "the exchange of goods and services, especially on a large scale." It includes legal, economic, political, social, cultural and technological systems that operate in any country or internationally.

:: International trade ::

A _____ is a document issued by a carrier to acknowledge receipt of cargo for shipment. Although the term historically related only to carriage by sea, a _____ may today be used for any type of carriage of goods.

Exam Probability: **Medium**

1. Answer choices:

(see index for correct answer)

- a. Trade
- b. Proexport
- c. Absolute advantage
- d. Bill of lading

Guidance: level 1

:: Economic globalization ::

_____ is an agreement in which one company hires another company to be responsible for a planned or existing activity that is or could be done internally, and sometimes involves transferring employees and assets from one firm to another.

Exam Probability: **High**

2. Answer choices:

(see index for correct answer)

- a. Outsourcing
- b. reshoring

Guidance: level 1

A _____ consists of one people who live in the same dwelling and share meals. It may also consist of a single family or another group of people. A dwelling is considered to contain multiple _____ s if meals or living spaces are not shared. The _____ is the basic unit of analysis in many social, microeconomic and government models, and is important to economics and inheritance.

Exam Probability: **Medium**

3. *Answer choices:*

(see index for correct answer)

- a. Household
- b. process perspective
- c. interpersonal communication
- d. personal values

Guidance: level 1

_____ is an emotion involving pleasure, , or anxiety in considering or awaiting an expected event.

Exam Probability: **High**

4. Answer choices:

(see index for correct answer)

- a. deep-level diversity
- b. Anticipation
- c. functional perspective
- d. co-culture

Guidance: level 1

:: Strategic alliances ::

A _____ is an agreement between two or more parties to pursue a set of agreed upon objectives needed while remaining independent organizations. A _____ will usually fall short of a legal partnership entity, agency, or corporate affiliate relationship. Typically, two companies form a _____ when each possesses one or more business assets or have expertise that will help the other by enhancing their businesses. _____ s can develop in outsourcing relationships where the parties desire to achieve long-term win-win benefits and innovation based on mutually desired outcomes.

Exam Probability: **Medium**

5. Answer choices:

(see index for correct answer)

- a. Strategic alliance
- b. International joint venture

- c. Cross-licensing
- d. Bridge Alliance

Guidance: level 1

:: ::

_____ is the process of removing or reducing state regulations, typically in the economic sphere. It is the repeal of governmental regulation of the economy. It became common in advanced industrial economies in the 1970s and 1980s, as a result of new trends in economic thinking about the inefficiencies of government regulation, and the risk that regulatory agencies would be controlled by the regulated industry to its benefit, and thereby hurt consumers and the wider economy.

Exam Probability: **Low**

6. *Answer choices:*

(see index for correct answer)

- a. information systems assessment
- b. process perspective
- c. Deregulation
- d. hierarchical

Guidance: level 1

:: ::

_____, or auditory perception, is the ability to perceive sounds by detecting vibrations, changes in the pressure of the surrounding medium through time, through an organ such as the ear. The academic field concerned with _____ is auditory science.

Exam Probability: **Low**

7. *Answer choices:*

(see index for correct answer)

- a. Hearing
- b. corporate values
- c. process perspective
- d. deep-level diversity

Guidance: level 1

:: ::

A _____ is a sworn body of people convened to render an impartial verdict officially submitted to them by a court, or to set a penalty or judgment. Modern juries tend to be found in courts to ascertain the guilt or lack thereof in a crime. In Anglophone jurisdictions, the verdict may be guilty or not guilty. The old institution of grand juries still exists in some places, particularly the United States, to investigate whether enough evidence of a crime exists to bring someone to trial.

Exam Probability: **High**

8. *Answer choices:*

(see index for correct answer)

- a. functional perspective
- b. Sarbanes-Oxley act of 2002
- c. open system
- d. Jury

Guidance: level 1

:: Investment ::

In finance, the benefit from an _____ is called a return. The return may consist of a gain realised from the sale of property or an _____ , unrealised capital appreciation , or _____ income such as dividends, interest, rental income etc., or a combination of capital gain and income. The return may also include currency gains or losses due to changes in foreign currency exchange rates.

Exam Probability: **Low**

9. *Answer choices:*

(see index for correct answer)

- a. Investment
- b. Multi-manager investment

- c. Passive investor
- d. Buy to let

Guidance: level 1

:: Management accounting ::

In economics, _____ s, indirect costs or overheads are business expenses that are not dependent on the level of goods or services produced by the business. They tend to be time-related, such as interest or rents being paid per month, and are often referred to as overhead costs. This is in contrast to variable costs, which are volume-related and unknown at the beginning of the accounting year. For a simple example, such as a bakery, the monthly rent for the baking facilities, and the monthly payments for the security system and basic phone line are _____ s, as they do not change according to how much bread the bakery produces and sells. On the other hand, the wage costs of the bakery are variable, as the bakery will have to hire more workers if the production of bread increases. Economists reckon _____ as a entry barrier for new entrepreneurs.

Exam Probability: **Low**

10. *Answer choices:*

(see index for correct answer)

- a. Fixed cost
- b. Construction accounting
- c. Spend management
- d. Certified Management Accountant

Guidance: level 1

:: Insolvency ::

_____ is the process in accounting by which a company is brought to an end in the United Kingdom, Republic of Ireland and United States. The assets and property of the company are redistributed. _____ is also sometimes referred to as winding-up or dissolution, although dissolution technically refers to the last stage of _____ . The process of _____ also arises when customs, an authority or agency in a country responsible for collecting and safeguarding customs duties, determines the final computation or ascertainment of the duties or drawback accruing on an entry.

Exam Probability: **High**

11. *Answer choices:*

(see index for correct answer)

- a. George Samuel Ford
- b. Bankruptcy
- c. Preferential creditor
- d. United Kingdom insolvency law

Guidance: level 1

:: Asset ::

In financial accounting, an _____ is any resource owned by the business. Anything tangible or intangible that can be owned or controlled to produce value and that is held by a company to produce positive economic value is an _____ . Simply stated, _____ s represent value of ownership that can be converted into cash . The balance sheet of a firm records the monetary value of the _____ s owned by that firm. It covers money and other valuables belonging to an individual or to a business.

Exam Probability: **Low**

12. *Answer choices:*

(see index for correct answer)

- a. Asset
- b. Fixed asset

Guidance: level 1

:: ::

_____ , also referred to as orthostasis, is a human position in which the body is held in an upright position and supported only by the feet.

Exam Probability: **High**

13. *Answer choices:*

(see index for correct answer)

- a. Standing
- b. hierarchical
- c. levels of analysis
- d. empathy

Guidance: level 1

:: International trade ::

_____ involves the transfer of goods or services from one person or entity to another, often in exchange for money. A system or network that allows _____ is called a market.

Exam Probability: **Medium**

14. *Answer choices:*

(see index for correct answer)

- a. Trade
- b. Trade commissioner
- c. Effective rate of protection
- d. Monetary hegemony

Guidance: level 1

:: Commercial item transport and distribution ::

A _____, forwarder, or forwarding agent, also known as a non-vessel operating common carrier, is a person or company that organizes shipments for individuals or corporations to get goods from the manufacturer or producer to a market, customer or final point of distribution. Forwarders contract with a carrier or often multiple carriers to move the goods. A forwarder does not move the goods but acts as an expert in the logistics network. These carriers can use a variety of shipping modes, including ships, airplanes, trucks, and railroads, and often do utilize multiple modes for a single shipment. For example, the _____ may arrange to have cargo moved from a plant to an airport by truck, flown to the destination city, then moved from the airport to a customer's building by another truck.

Exam Probability: **Medium**

15. *Answer choices:*

(see index for correct answer)

- a. Cargo sampling
- b. Ship transport
- c. Freight forwarder
- d. Containerlift

Guidance: level 1

:: Information retrieval ::

_____ is a technique used by recommender systems. _____ has two senses, a narrow one and a more general one.

Exam Probability: **Low**

16. *Answer choices:*

(see index for correct answer)

- a. Legal information retrieval
- b. Communication engine
- c. Isearch
- d. Temporal information retrieval

Guidance: level 1

:: ::

A _____ or _____ s is a type of footwear and not a specific type of shoe. Most _____ s mainly cover the foot and the ankle, while some also cover some part of the lower calf. Some _____ s extend up the leg, sometimes as far as the knee or even the hip. Most _____ s have a heel that is clearly distinguishable from the rest of the sole, even if the two are made of one piece. Traditionally made of leather or rubber, modern _____ s are made from a variety of materials. _____ s are worn both for their functionality protecting the foot and leg from water, extreme cold, mud or hazards or providing additional ankle support for strenuous activities with added traction requirements , or may have hobnails on their undersides to protect against wear and to get better grip; and for reasons of style and fashion.

Exam Probability: **Low**

17. *Answer choices:*

(see index for correct answer)

- a. functional perspective
- b. levels of analysis
- c. Boot
- d. cultural

Guidance: level 1

:: Basic financial concepts ::

_____ is a sustained increase in the general price level of goods and services in an economy over a period of time. When the general price level rises, each unit of currency buys fewer goods and services; consequently, _____ reflects a reduction in the purchasing power per unit of money a loss of real value in the medium of exchange and unit of account within the economy. The measure of _____ is the _____ rate, the annualized percentage change in a general price index, usually the consumer price index, over time. The opposite of _____ is deflation.

Exam Probability: **High**

18. *Answer choices:*

(see index for correct answer)

- a. Short interest
- b. Future-oriented

- c. Eurodollar
- d. Financial transaction

Guidance: level 1

:: Theories ::

A _____ union is a type of multinational political union where negotiated power is delegated to an authority by governments of member states.

Exam Probability: **High**

19. *Answer choices:*
(see index for correct answer)

- a. incrementalism
- b. Supranational

Guidance: level 1

:: Cryptography ::

In cryptography, _____ is the process of encoding a message or information in such a way that only authorized parties can access it and those who are not authorized cannot. _____ does not itself prevent interference, but denies the intelligible content to a would-be interceptor. In an _____ scheme, the intended information or message, referred to as plaintext, is encrypted using an _____ algorithm – a cipher – generating ciphertext that can be read only if decrypted. For technical reasons, an _____ scheme usually uses a pseudo-random _____ key generated by an algorithm. It is in principle possible to decrypt the message without possessing the key, but, for a well-designed _____ scheme, considerable computational resources and skills are required. An authorized recipient can easily decrypt the message with the key provided by the originator to recipients but not to unauthorized users.

Exam Probability: **High**

20. *Answer choices:*

(see index for correct answer)

- a. plaintext
- b. Anonymous matching
- c. ciphertext
- d. Electronic Signature

Guidance: level 1

:: Budgets ::

A _____ is a financial plan for a defined period, often one year. It may also include planned sales volumes and revenues, resource quantities, costs and expenses, assets, liabilities and cash flows. Companies, governments, families and other organizations use it to express strategic plans of activities or events in measurable terms.

Exam Probability: **Medium**

21. *Answer choices:*

(see index for correct answer)

- a. Performance-based budgeting
- b. Participatory budgeting
- c. Railway Budget
- d. Envelope system

Guidance: level 1

:: Business terms ::

_____ning is an organization's process of defining its strategy, or direction, and making decisions on allocating its resources to pursue this strategy. It may also extend to control mechanisms for guiding the implementation of the strategy. _____ning became prominent in corporations during the 1960s and remains an important aspect of strategic management. It is executed by _____ners or strategists, who involve many parties and research sources in their analysis of the organization and its relationship to the environment in which it competes.

Exam Probability: **Low**

22. *Answer choices:*

(see index for correct answer)

- a. Strategic plan
- b. churn rate
- c. granular
- d. back office

Guidance: level 1

:: E-commerce ::

_____ is the business-to-business or business-to-consumer or business-to-government purchase and sale of supplies, work, and services through the Internet as well as other information and networking systems, such as electronic data interchange and enterprise resource planning.

Exam Probability: **High**

23. *Answer choices:*

(see index for correct answer)

- a. E-procurement
- b. Webjet
- c. AbleCommerce

- d. Global Location Number

Guidance: level 1

:: Commerce ::

_____, Inc. is an American media-services provider headquartered in Los Gatos, California, founded in 1997 by Reed Hastings and Marc Randolph in Scotts Valley, California. The company's primary business is its subscription-based streaming OTT service which offers online streaming of a library of films and television programs, including those produced in-house. As of April 2019, _____ had over 148 million paid subscriptions worldwide, including 60 million in the United States, and over 154 million subscriptions total including free trials. It is available almost worldwide except in mainland China as well as Syria, North Korea, and Crimea. The company also has offices in the Netherlands, Brazil, India, Japan, and South Korea. _____ is a member of the Motion Picture Association of America.

Exam Probability: **Low**

24. *Answer choices:*
(see index for correct answer)

- a. Netflix
- b. Hong Kong Mercantile Exchange
- c. Closed household economy
- d. Trading post

Guidance: level 1

:: Minimum wage ::

A _____ is the lowest remuneration that employers can legally pay their workers—the price floor below which workers may not sell their labor. Most countries had introduced _____ legislation by the end of the 20th century.

Exam Probability: **Low**

25. *Answer choices:*

(see index for correct answer)

- a. Guaranteed minimum income
- b. Minimum Wage Fairness Act
- c. Minimum wage in the United States
- d. National Anti-Sweating League

Guidance: level 1

:: ::

Employment is a relationship between two parties, usually based on a contract where work is paid for, where one party, which may be a corporation, for profit, not-for-profit organization, co-operative or other entity is the employer and the other is the employee. Employees work in return for payment, which may be in the form of an hourly wage, by piecework or an annual salary, depending on the type of work an employee does or which sector she or he is working in. Employees in some fields or sectors may receive gratuities, bonus payment or stock options. In some types of employment, employees may receive benefits in addition to payment. Benefits can include health insurance, housing, disability insurance or use of a gym. Employment is typically governed by employment laws, regulations or legal contracts.

Exam Probability: **Medium**

26. *Answer choices:*

(see index for correct answer)

- a. imperative
- b. co-culture
- c. Personnel
- d. information systems assessment

Guidance: level 1

:: ::

_____ refers to a business or organization attempting to acquire goods or services to accomplish its goals. Although there are several organizations that attempt to set standards in the _____ process, processes can vary greatly between organizations. Typically the word "_____" is not used interchangeably with the word "procurement", since procurement typically includes expediting, supplier quality, and transportation and logistics in addition to _____ .

Exam Probability: **Medium**

27. *Answer choices:*

(see index for correct answer)

- a. corporate values
- b. information systems assessment
- c. hierarchical perspective
- d. levels of analysis

Guidance: level 1

:: E-commerce ::

_____ , cybersecurity or information technology security is the protection of computer systems from theft or damage to their hardware, software or electronic data, as well as from disruption or misdirection of the services they provide.

Exam Probability: **Medium**

28. Answer choices:

(see index for correct answer)

- a. Online locator service
- b. Urban Ladder
- c. Customer Access and Retrieval System
- d. Direct imports

Guidance: level 1

:: Commercial item transport and distribution ::

Wholesaling or distributing is the sale of goods or merchandise to retailers; to industrial, commercial, institutional, or other professional business users; or to other _____rs and related subordinated services. In general, it is the sale of goods to anyone other than a standard consumer.

Exam Probability: **Medium**

29. Answer choices:

(see index for correct answer)

- a. LNG carrier
- b. Warehouse receipt
- c. Australia standard pallets
- d. Wholesale

Guidance: level 1

:: Dot-com bubble ::

_____ is an internet portal launched in 1995 that provides a variety of content including news and weather, a metasearch engine, a web-based email, instant messaging, stock quotes, and a customizable user homepage. It is currently operated by IAC Applications of IAC, and _____ Networks. In the U.S., the main _____ site has long been a personal start page called My _____ . _____ also operates an e-mail service, although it is no longer open for new customers.

Exam Probability: **Medium**

30. *Answer choices:*
(see index for correct answer)

- a. CyberRebate
- b. GeoCities
- c. Cyberian Outpost
- d. The Industry Standard

Guidance: level 1

:: Management ::

In business, a _____ is the attribute that allows an organization to outperform its competitors. A _____ may include access to natural resources, such as high-grade ores or a low-cost power source, highly skilled labor, geographic location, high entry barriers, and access to new technology.

Exam Probability: **Medium**

31. *Answer choices:*

(see index for correct answer)

- a. Competitive advantage
- b. Total Worker Health
- c. Quality, cost, delivery
- d. Supply network

Guidance: level 1

:: Free market ::

In economics, a _____ is a system in which the prices for goods and services are determined by the open market and by consumers. In a _____, the laws and forces of supply and demand are free from any intervention by a government or other authority and from all forms of economic privilege, monopolies and artificial scarcities. Proponents of the concept of _____ contrast it with a regulated market in which a government intervenes in supply and demand through various methods, such as tariffs, used to restrict trade and to protect the local economy. In an idealized free-market economy, prices for goods and services are set freely by the forces of supply and demand and are allowed to reach their point of equilibrium without intervention by government policy.

Exam Probability: **Low**

32. *Answer choices:*
(see index for correct answer)

- a. Piece rate
- b. Regulated market

Guidance: level 1

:: Human resource management ::

_____ are the people who make up the workforce of an organization, business sector, or economy. "Human capital" is sometimes used synonymously with " _____ ", although human capital typically refers to a narrower effect. Likewise, other terms sometimes used include manpower, talent, labor, personnel, or simply people.

Exam Probability: **High**

33. *Answer choices:*

(see index for correct answer)

- a. Trust fall
- b. Lego Serious Play
- c. Organizational orientations
- d. Human resources

Guidance: level 1

:: ::

_____ is the study and management of exchange relationships. _____ is the business process of creating relationships with and satisfying customers. With its focus on the customer, _____ is one of the premier components of business management.

Exam Probability: **Low**

34. *Answer choices:*

(see index for correct answer)

- a. interpersonal communication
- b. co-culture
- c. imperative

- d. personal values

Guidance: level 1

:: Organizational structure ::

An _____ defines how activities such as task allocation, coordination, and supervision are directed toward the achievement of organizational aims.

Exam Probability: **High**

35. *Answer choices:*
(see index for correct answer)

- a. Organization of the New York City Police Department
- b. Automated Bureaucracy
- c. Blessed Unrest
- d. Organizational structure

Guidance: level 1

:: Stock market ::

The _____ of a corporation is all of the shares into which ownership of the corporation is divided. In American English, the shares are commonly known as "_____ s". A single share of the _____ represents fractional ownership of the corporation in proportion to the total number of shares. This typically entitles the _____ holder to that fraction of the company's earnings, proceeds from liquidation of assets, or voting power, often dividing these up in proportion to the amount of money each _____ holder has invested. Not all _____ is necessarily equal, as certain classes of _____ may be issued for example without voting rights, with enhanced voting rights, or with a certain priority to receive profits or liquidation proceeds before or after other classes of shareholders.

Exam Probability: **High**

36. *Answer choices:*

(see index for correct answer)

- a. Leading stock
- b. Abnormal return
- c. Central securities depository
- d. CNBC World

Guidance: level 1

:: ::

A _____ is a person or firm who arranges transactions between a buyer and a seller for a commission when the deal is executed. A _____ who also acts as a seller or as a buyer becomes a principal party to the deal. Neither role should be confused with that of an agent—one who acts on behalf of a principal party in a deal.

Exam Probability: **High**

37. *Answer choices:*

(see index for correct answer)

- a. Broker
- b. hierarchical perspective
- c. interpersonal communication
- d. deep-level diversity

Guidance: level 1

:: Marketing ::

_____ —an information- and communication-based electronic exchange environment—is a relatively new concept in marketing. Since physical boundaries no longer interfere with buy/sell decisions, the world has grown into several industry specific _____ s which are integration of marketplaces through sophisticated computer and telecommunication technologies. The term _____ was introduced by Jeffrey Rayport and John Sviokla in 1994 in their article "Managing in the _____ " that appeared in Harvard Business Review. In the article the authors distinguished between electronic and conventional markets. In a _____ , information and/or physical goods are exchanged, and transactions take place through computers and networks. These networks consist of blogs, forum threads, and micro-blogging services like Twitter. Businesses and their customers are enabled to create conversations and two-way communications about products and services. These conversations may also happen outside the sphere of control of a given business, when a marketing campaign or customer-service issue captures the attention of web-savvy consumers.

Exam Probability: **Medium**

38. *Answer choices:*

(see index for correct answer)

- a. Adobe Marketing Cloud
- b. Marketspace
- c. Macromarketing
- d. Bluetooth advertising

Guidance: level 1

:: Information technology ::

_____ is the use of computers to store, retrieve, transmit, and manipulate data, or information, often in the context of a business or other enterprise. IT is considered to be a subset of information and communications technology. An _____ system is generally an information system, a communications system or, more specifically speaking, a computer system – including all hardware, software and peripheral equipment – operated by a limited group of users.

Exam Probability: **Medium**

39. *Answer choices:*

(see index for correct answer)

- a. CIO-plus
- b. Enumerate
- c. Information technology
- d. Qualitest Group

Guidance: level 1

:: Commodities ::

In economics, a _____ is an economic good or service that has full or substantial fungibility: that is, the market treats instances of the good as equivalent or nearly so with no regard to who produced them. Most commodities are raw materials, basic resources, agricultural, or mining products, such as iron ore, sugar, or grains like rice and wheat. Commodities can also be mass-produced unspecialized products such as chemicals and computer memory.

Exam Probability: **Medium**

40. *Answer choices:*

(see index for correct answer)

- a. Sample grade
- b. IRely
- c. Commodity pathway diversion
- d. Commodity

Guidance: level 1

:: Production economics ::

In economics long run is a theoretical concept where all markets are in equilibrium, and all prices and quantities have fully adjusted and are in equilibrium. The long run contrasts with the _____ where there are some constraints and markets are not fully in equilibrium.

Exam Probability: **High**

41. *Answer choices:*

(see index for correct answer)

- a. Transaction cost
- b. Productive capacity
- c. The labor problem

- d. Short run

Guidance: level 1

:: E-commerce ::

An _____ , or automated clearinghouse, is an electronic network for financial transactions, generally domestic low value payments. An ACH is a computer-based clearing house and settlement facility established to process the exchange of electronic transactions between participating financial institutions. It is a form of clearing house that is specifically for payments and may support both credit transfers and direct debits.

Exam Probability: **High**

42. *Answer choices:*

(see index for correct answer)

- a. FabMart
- b. Automated Clearing House
- c. Helpling
- d. Variable pricing

Guidance: level 1

:: Supply chain management ::

A _____ is a type of auction in which the traditional roles of buyer and seller are reversed. Thus, there is one buyer and many potential sellers. In an ordinary auction, buyers compete to obtain goods or services by offering increasingly higher prices. In contrast, in a _____, the sellers compete to obtain business from the buyer and prices will typically decrease as the sellers underbid each other.

Exam Probability: **High**

43. *Answer choices:*
(see index for correct answer)

- a. Universal Product Code
- b. Reverse auction
- c. Reverse logistics
- d. ICON-SCM

Guidance: level 1

:: ::

_____ is a means of protection from financial loss. It is a form of risk management, primarily used to hedge against the risk of a contingent or uncertain loss

Exam Probability: **High**

44. Answer choices:

(see index for correct answer)

- a. personal values
- b. hierarchical perspective
- c. levels of analysis
- d. Insurance

Guidance: level 1

:: International trade ::

An _____ is a good brought into a jurisdiction, especially across a national border, from an external source. The party bringing in the good is called an _____ er. An _____ in the receiving country is an export from the sending country. _____ ation and exportation are the defining financial transactions of international trade.

Exam Probability: **High**

45. Answer choices:

(see index for correct answer)

- a. AgroEurasia
- b. Foreign affiliate trade statistics
- c. FAST Card
- d. European Customs Information Portal

:: Commerce ::

An _____ is a bank that offers card association branded payment cards directly to consumers. The name is derived from the practice of issuing payment to the acquiring bank on behalf of its customer.

Exam Probability: **High**

46. *Answer choices:*

(see index for correct answer)

- a. Retail loss prevention
- b. Issuing bank
- c. Weight
- d. Bargaining power

In a supply chain, a _____, or a seller, is an enterprise that contributes goods or services. Generally, a supply chain _____ manufactures inventory/stock items and sells them to the next link in the chain. Today, these terms refer to a supplier of any good or service.

Exam Probability: **Medium**

47. *Answer choices:*

(see index for correct answer)

- a. personal values
- b. process perspective
- c. Vendor
- d. interpersonal communication

Guidance: level 1

:: ::

_____ is the amount of time someone works beyond normal working hours. The term is also used for the pay received for this time. Normal hours may be determined in several ways.

Exam Probability: **Medium**

48. *Answer choices:*

(see index for correct answer)

- a. process perspective
- b. Overtime
- c. interpersonal communication
- d. hierarchical perspective

Guidance: level 1

:: ::

According to the philosopher Piyush Mathur, "Tangibility is the property that a phenomenon exhibits if it has and/or transports mass and/or energy and/or momentum".

Exam Probability: **Low**

49. *Answer choices:*

(see index for correct answer)

- a. open system
- b. Tangible
- c. Sarbanes-Oxley act of 2002
- d. imperative

Guidance: level 1

:: Auctioneering ::

An _____ is a process of buying and selling goods or services by offering them up for bid, taking bids, and then selling the item to the highest bidder. The open ascending price _____ is arguably the most common form of _____ in use today. Participants bid openly against one another, with each subsequent bid required to be higher than the previous bid. An _____ eer may announce prices, bidders may call out their bids themselves, or bids may be submitted electronically with the highest current bid publicly displayed. In a Dutch _____, the _____ eer begins with a high asking price for some quantity of like items; the price is lowered until a participant is willing to accept the _____ eer's price for some quantity of the goods in the lot or until the seller's reserve price is met. While _____ s are most associated in the public imagination with the sale of antiques, paintings, rare collectibles and expensive wines, _____ s are also used for commodities, livestock, radio spectrum and used cars. In economic theory, an _____ may refer to any mechanism or set of trading rules for exchange.

Exam Probability: **Low**

50. *Answer choices:*

(see index for correct answer)

- a. Art auction
- b. Japanese auction
- c. English auction
- d. Auction

Guidance: level 1

:: Confidence tricks ::

_____ is the fraudulent attempt to obtain sensitive information such as usernames, passwords and credit card details by disguising oneself as a trustworthy entity in an electronic communication. Typically carried out by email spoofing or instant messaging, it often directs users to enter personal information at a fake website which matches the look and feel of the legitimate site.

Exam Probability: **Low**

51. *Answer choices:*

(see index for correct answer)

- a. Private investment capital subscription
- b. Hokkani boro
- c. The switch
- d. Technical support scam

Guidance: level 1

:: ::

_____ is a term frequently used in marketing. It is a measure of how products and services supplied by a company meet or surpass customer expectation. _____ is defined as "the number of customers, or percentage of total customers, whose reported experience with a firm, its products, or its services exceeds specified satisfaction goals."

Exam Probability: **Low**

52. *Answer choices:*

(see index for correct answer)

- a. similarity-attraction theory
- b. Sarbanes-Oxley act of 2002
- c. Customer satisfaction
- d. deep-level diversity

Guidance: level 1

:: Debt ::

_____ , in finance and economics, is payment from a borrower or deposit-taking financial institution to a lender or depositor of an amount above repayment of the principal sum , at a particular rate. It is distinct from a fee which the borrower may pay the lender or some third party. It is also distinct from dividend which is paid by a company to its shareholders from its profit or reserve, but not at a particular rate decided beforehand, rather on a pro rata basis as a share in the reward gained by risk taking entrepreneurs when the revenue earned exceeds the total costs.

Exam Probability: **Medium**

53. *Answer choices:*

(see index for correct answer)

- a. Odious debt
- b. Internal debt
- c. Interest
- d. Debt

Guidance: level 1

:: Management ::

The term _____ refers to measures designed to increase the degree of autonomy and self-determination in people and in communities in order to enable them to represent their interests in a responsible and self-determined way, acting on their own authority. It is the process of becoming stronger and more confident, especially in controlling one's life and claiming one's rights. _____ as action refers both to the process of self- _____ and to professional support of people, which enables them to overcome their sense of powerlessness and lack of influence, and to recognize and use their resources. To do work with power.

Exam Probability: **Medium**

54. *Answer choices:*

(see index for correct answer)

- a. Corporate foresight
- b. Performance indicator
- c. Energy management software
- d. Empowerment

Guidance: level 1

:: Goods ::

In most contexts, the concept of _____ denotes the conduct that should be preferred when posed with a choice between possible actions. _____ is generally considered to be the opposite of evil, and is of interest in the study of morality, ethics, religion and philosophy. The specific meaning and etymology of the term and its associated translations among ancient and contemporary languages show substantial variation in its inflection and meaning depending on circumstances of place, history, religious, or philosophical context.

Exam Probability: **Medium**

55. *Answer choices:*

(see index for correct answer)

- a. Information good
- b. Giffen good
- c. Good
- d. Bad

Guidance: level 1

:: Marketing techniques ::

_____ is the activity of dividing a broad consumer or business market, normally consisting of existing and potential customers, into sub-groups of consumers based on some type of shared characteristics. In dividing or segmenting markets, researchers typically look for common characteristics such as shared needs, common interests, similar lifestyles or even similar demographic profiles. The overall aim of segmentation is to identify high yield segments – that is, those segments that are likely to be the most profitable or that have growth potential – so that these can be selected for special attention.

Exam Probability: **Low**

56. *Answer choices:*

(see index for correct answer)

- a. Fear, uncertainty and doubt
- b. Elevator pitch
- c. Market segmentation
- d. Loss leader

Guidance: level 1

:: Marketing analytics ::

_____ is a long-term, forward-looking approach to planning with the fundamental goal of achieving a sustainable competitive advantage. Strategic planning involves an analysis of the company's strategic initial situation prior to the formulation, evaluation and selection of market-oriented competitive position that contributes to the company's goals and marketing objectives.

Exam Probability: **Medium**

57. *Answer choices:*

(see index for correct answer)

- a. marketing dashboard
- b. Advertising adstock
- c. Marketing strategy
- d. Sumall

Guidance: level 1

:: Business law ::

A _____ is a contractual arrangement calling for the lessee to pay the lessor for use of an asset. Property, buildings and vehicles are common assets that are _____ d. Industrial or business equipment is also _____ d.

Exam Probability: **Low**

58. Answer choices:

(see index for correct answer)

- a. Lease
- b. Leave of absence
- c. Law of agency
- d. Negative option billing

Guidance: level 1

:: Commercial item transport and distribution ::

In a contract of carriage, the _____ is the entity who is financially responsible for the receipt of a shipment. Generally, but not always, the _____ is the same as the receiver.

Exam Probability: **Low**

59. Answer choices:

(see index for correct answer)

- a. Common carrier
- b. EPCglobal Network
- c. Food distribution
- d. Containerized housing unit

Guidance: level 1

Business ethics

Business ethics (also known as corporate ethics) is a form of applied ethics or professional ethics, that examines ethical principles and moral or ethical problems that can arise in a business environment. It applies to all aspects of business conduct and is relevant to the conduct of individuals and entire organizations. These ethics originate from individuals, organizational statements or from the legal system. These norms, values, ethical, and unethical practices are what is used to guide business. They help those businesses maintain a better connection with their stakeholders.

The _____ was a severe worldwide economic depression that took place mostly during the 1930s, beginning in the United States. The timing of the _____ varied across nations; in most countries it started in 1929 and lasted until the late-1930s. It was the longest, deepest, and most widespread depression of the 20th century. In the 21st century, the _____ is commonly used as an example of how intensely the world's economy can decline.

Exam Probability: **Medium**

1. *Answer choices:*

(see index for correct answer)

- a. open system
- b. Great Depression
- c. Sarbanes-Oxley act of 2002
- d. similarity-attraction theory

Guidance: level 1

:: United States law ::

The ABA _____ , created by the American Bar Association , are a set of rules that prescribe baseline standards of legal ethics and professional responsibility for lawyers in the United States. They were promulgated by the ABA House of Delegates upon the recommendation of the Kutak Commission in 1983. The rules are merely recommendations, or models, and are not themselves binding. However, having a common set of Model Rules facilitates a common discourse on legal ethics, and simplifies professional responsibility training as well as the day-to-day application of such rules. As of 2015, 49 states and four territories have adopted the rules in whole or in part, of which the most recent to do so was the Commonwealth of the Northern Mariana Islands in March 2015. California is the only state that has not adopted the ABA Model Rules, while Puerto Rico is the only U.S. jurisdiction outside of confederation has not adopted them but instead has its own Código de Ética Profesional.

Exam Probability: **Low**

2. *Answer choices:*

(see index for correct answer)

- a. judgment notwithstanding the verdict
- b. Model Rules of Professional Conduct

Guidance: level 1

:: Leadership ::

_____ is a theory of leadership where a leader works with teams to identify needed change, creating a vision to guide the change through inspiration, and executing the change in tandem with committed members of a group; it is an integral part of the Full Range Leadership Model. _____ serves to enhance the motivation, morale, and job performance of followers through a variety of mechanisms; these include connecting the follower's sense of identity and self to a project and to the collective identity of the organization; being a role model for followers in order to inspire them and to raise their interest in the project; challenging followers to take greater ownership for their work, and understanding the strengths and weaknesses of followers, allowing the leader to align followers with tasks that enhance their performance.

Exam Probability: **High**

3. *Answer choices:*

(see index for correct answer)

- a. Meta-leadership
- b. Transactional leadership
- c. Transformational leadership
- d. Situational leadership

Guidance: level 1

_____ was a philosopher during the Classical period in Ancient Greece, the founder of the Lyceum and the Peripatetic school of philosophy and Aristotelian tradition. Along with his teacher Plato, he is considered the "Father of Western Philosophy". His writings cover many subjects – including physics, biology, zoology, metaphysics, logic, ethics, aesthetics, poetry, theatre, music, rhetoric, psychology, linguistics, economics, politics and government. _____ provided a complex synthesis of the various philosophies existing prior to him, and it was above all from his teachings that the West inherited its intellectual lexicon, as well as problems and methods of inquiry. As a result, his philosophy has exerted a unique influence on almost every form of knowledge in the West and it continues to be a subject of contemporary philosophical discussion.

Exam Probability: **Medium**

4. *Answer choices:*

(see index for correct answer)

- a. levels of analysis
- b. Aristotle
- c. corporate values
- d. similarity-attraction theory

Guidance: level 1

:: Production and manufacturing ::

_____ is a set of techniques and tools for process improvement. Though as a shortened form it may be found written as 6S, it should not be confused with the methodology known as 6S.

Exam Probability: **High**

5. *Answer choices:*

(see index for correct answer)

- a. Value engineering
- b. ISO/IEC 17025
- c. Six Sigma
- d. Craft production

Guidance: level 1

:: ::

Competition law is a law that promotes or seeks to maintain market competition by regulating anti-competitive conduct by companies. Competition law is implemented through public and private enforcement. Competition law is known as "_____ law" in the United States for historical reasons, and as "anti-monopoly law" in China and Russia. In previous years it has been known as trade practices law in the United Kingdom and Australia. In the European Union, it is referred to as both _____ and competition law.

Exam Probability: **Low**

6. *Answer choices:*

(see index for correct answer)

- a. deep-level diversity
- b. Antitrust
- c. co-culture
- d. open system

Guidance: level 1

:: Water law ::

The _____ is the primary federal law in the United States governing water pollution. Its objective is to restore and maintain the chemical, physical, and biological integrity of the nation's waters; recognizing the responsibilities of the states in addressing pollution and providing assistance to states to do so, including funding for publicly owned treatment works for the improvement of wastewater treatment; and maintaining the integrity of wetlands. It is one of the United States' first and most influential modern environmental laws. As with many other major U.S. federal environmental statutes, it is administered by the U.S. Environmental Protection Agency , in coordination with state governments. Its implementing regulations are codified at 40 C.F.R. Subchapters D, N, and O .

Exam Probability: **High**

7. *Answer choices:*

(see index for correct answer)

- a. The Helsinki Rules on the Uses of the Waters of International Rivers
- b. Correlative rights doctrine
- c. Water right
- d. Clean Water Act

Guidance: level 1

:: United Kingdom labour law ::

The _____ was a series of programs, public work projects, financial reforms, and regulations enacted by President Franklin D. Roosevelt in the United States between 1933 and 1936. It responded to needs for relief, reform, and recovery from the Great Depression. Major federal programs included the Civilian Conservation Corps , the Civil Works Administration , the Farm Security Administration , the National Industrial Recovery Act of 1933 and the Social Security Administration . They provided support for farmers, the unemployed, youth and the elderly. The _____ included new constraints and safeguards on the banking industry and efforts to re-inflate the economy after prices had fallen sharply. _____ programs included both laws passed by Congress as well as presidential executive orders during the first term of the presidency of Franklin D. Roosevelt.

Exam Probability: **Low**

8. *Answer choices:*

(see index for correct answer)

- a. Collective Redundancies Directive
- b. Industrial Law Journal

- c. Collective action in the United Kingdom
- d. New Deal

Guidance: level 1

:: Confidence tricks ::

A _____ is a business model that recruits members via a promise of payments or services for enrolling others into the scheme, rather than supplying investments or sale of products. As recruiting multiplies, recruiting becomes quickly impossible, and most members are unable to profit; as such, _____ s are unsustainable and often illegal.

Exam Probability: **Low**

9. *Answer choices:*

(see index for correct answer)

- a. Moving scam
- b. Patent safe
- c. The switch
- d. Private investment capital subscription

Guidance: level 1

:: Business ethics ::

_____ is a persistent pattern of mistreatment from others in the workplace that causes either physical or emotional harm. It can include such tactics as verbal, nonverbal, psychological, physical abuse and humiliation. This type of workplace aggression is particularly difficult because, unlike the typical school bully, workplace bullies often operate within the established rules and policies of their organization and their society. In the majority of cases, bullying in the workplace is reported as having been by someone who has authority over their victim. However, bullies can also be peers, and occasionally subordinates. Research has also investigated the impact of the larger organizational context on bullying as well as the group-level processes that impact on the incidence and maintenance of bullying behaviour. Bullying can be covert or overt. It may be missed by superiors; it may be known by many throughout the organization. Negative effects are not limited to the targeted individuals, and may lead to a decline in employee morale and a change in organizational culture. It can also take place as overbearing supervision, constant criticism, and blocking promotions.

Exam Probability: **High**

10. *Answer choices:*

(see index for correct answer)

- a. Interfaith Center on Corporate Responsibility
- b. Ethical corporate social responsibility
- c. Workplace bullying
- d. Contingent work

Guidance: level 1

:: Utilitarianism ::

_____ is a school of thought that argues that the pursuit of pleasure and intrinsic goods are the primary or most important goals of human life. A hedonist strives to maximize net pleasure. However upon finally gaining said pleasure, happiness may remain stationary.

Exam Probability: **Low**

11. *Answer choices:*

(see index for correct answer)

- a. Hedonism
- b. Utilitarian bioethics
- c. Mohism
- d. Equal consideration of interests

Guidance: level 1

:: Management ::

The term _____ refers to measures designed to increase the degree of autonomy and self-determination in people and in communities in order to enable them to represent their interests in a responsible and self-determined way, acting on their own authority. It is the process of becoming stronger and more confident, especially in controlling one's life and claiming one's rights. _____ as action refers both to the process of self-_____ and to professional support of people, which enables them to overcome their sense of powerlessness and lack of influence, and to recognize and use their resources. To do work with power.

Exam Probability: **Low**

12. *Answer choices:*

(see index for correct answer)

- a. Demand chain management
- b. Just in time
- c. Empowerment
- d. Supply management

Guidance: level 1

:: Social enterprise ::

Corporate social responsibility is a type of international private business self-regulation. While once it was possible to describe CSR as an internal organisational policy or a corporate ethic strategy, that time has passed as various international laws have been developed and various organisations have used their authority to push it beyond individual or even industry-wide initiatives. While it has been considered a form of corporate self-regulation for some time, over the last decade or so it has moved considerably from voluntary decisions at the level of individual organisations, to mandatory schemes at regional, national and even transnational levels.

Exam Probability: **Low**

13. *Answer choices:*

(see index for correct answer)

- a. Social enterprise
- b. Corporate citizenship

Guidance: level 1

:: ::

_____ is a bundle of characteristics, including ways of thinking, feeling, and acting, which humans are said to have naturally. The term is often regarded as capturing what it is to be human, or the essence of humanity. The term is controversial because it is disputed whether or not such an essence exists. Arguments about _____ have been a mainstay of philosophy for centuries and the concept continues to provoke lively philosophical debate. The concept also continues to play a role in science, with neuroscientists, psychologists and social scientists sometimes claiming that their results have yielded insight into _____ . _____ is traditionally contrasted with characteristics that vary among humans, such as characteristics associated with specific cultures. Debates about _____ are related to, although not the same as, debates about the comparative importance of genes and environment in development .

Exam Probability: **Low**

14. *Answer choices:*

(see index for correct answer)

- a. cultural
- b. Human nature
- c. empathy
- d. surface-level diversity

Guidance: level 1

:: Reputation management ::

_____ or image of a social entity is an opinion about that entity, typically as a result of social evaluation on a set of criteria.

Exam Probability: **Low**

15. *Answer choices:*
(see index for correct answer)

- a. Conversocial
- b. personal brand
- c. Reputation
- d. Raph Levien

Guidance: level 1

:: Financial markets ::

The _____ is a United States federal government organization, established by Title I of the Dodd–Frank Wall Street Reform and Consumer Protection Act, which was signed into law by President Barack Obama on July 21, 2010. The Office of Financial Research is intended to provide support to the council.

Exam Probability: **Medium**

16. *Answer choices:*

(see index for correct answer)

- a. Financial Stability Oversight Council
- b. Advanced Computerized Execution System
- c. Financial market
- d. Systematic trading

Guidance: level 1

:: ::

_____ is "property consisting of land and the buildings on it, along with its natural resources such as crops, minerals or water; immovable property of this nature; an interest vested in this an item of real property, buildings or housing in general. Also: the business of _____ ; the profession of buying, selling, or renting land, buildings, or housing." It is a legal term used in jurisdictions whose legal system is derived from English common law, such as India, England, Wales, Northern Ireland, United States, Canada, Pakistan, Australia, and New Zealand.

Exam Probability: **Low**

17. *Answer choices:*

(see index for correct answer)

- a. personal values
- b. hierarchical
- c. empathy
- d. interpersonal communication

Guidance: level 1

:: Environmental economics ::

_____ is the process of people maintaining change in a balanced environment, in which the exploitation of resources, the direction of investments, the orientation of technological development and institutional change are all in harmony and enhance both current and future potential to meet human needs and aspirations. For many in the field, _____ is defined through the following interconnected domains or pillars: environment, economic and social, which according to Fritjof Capra is based on the principles of Systems Thinking. Sub-domains of sustainable development have been considered also: cultural, technological and political. While sustainable development may be the organizing principle for _____ for some, for others, the two terms are paradoxical . Sustainable development is the development that meets the needs of the present without compromising the ability of future generations to meet their own needs. Brundtland Report for the World Commission on Environment and Development introduced the term of sustainable development.

Exam Probability: **Medium**

18. Answer choices:

(see index for correct answer)

- a. Gas depletion
- b. Assigned amount units
- c. Hot stain
- d. Sustainability

Guidance: level 1

:: Renewable energy ::

_____ is the conversion of energy from sunlight into electricity, either directly using photovoltaics, indirectly using concentrated _____, or a combination. Concentrated _____ systems use lenses or mirrors and tracking systems to focus a large area of sunlight into a small beam. Photovoltaic cells convert light into an electric current using the photovoltaic effect.

Exam Probability: **Low**

19. Answer choices:

(see index for correct answer)

- a. Solar power
- b. Deep water source cooling
- c. Yield co
- d. IRENA Renewable Energy Learning Partnership

Guidance: level 1

:: Progressive Era in the United States ::

The Clayton Antitrust Act of 1914, was a part of United States antitrust law with the goal of adding further substance to the U.S. antitrust law regime; the _____ sought to prevent anticompetitive practices in their incipiency. That regime started with the Sherman Antitrust Act of 1890, the first Federal law outlawing practices considered harmful to consumers. The _____ specified particular prohibited conduct, the three-level enforcement scheme, the exemptions, and the remedial measures.

Exam Probability: **Low**

20. *Answer choices:*
(see index for correct answer)

- a. pragmatism
- b. Mann Act
- c. Clayton Act

Guidance: level 1

:: ::

An _____ is the release of a liquid petroleum hydrocarbon into the environment, especially the marine ecosystem, due to human activity, and is a form of pollution. The term is usually given to marine _____ s, where oil is released into the ocean or coastal waters, but spills may also occur on land. _____ s may be due to releases of crude oil from tankers, offshore platforms, drilling rigs and wells, as well as spills of refined petroleum products and their by-products, heavier fuels used by large ships such as bunker fuel, or the spill of any oily refuse or waste oil.

Exam Probability: **Low**

21. *Answer choices:*

(see index for correct answer)

- a. imperative
- b. Sarbanes-Oxley act of 2002
- c. Character
- d. information systems assessment

Guidance: level 1

:: Pyramid and Ponzi schemes ::

_____ was an Italian swindler and con artist in the U.S. and Canada. His aliases include Charles Ponci, Carlo, and Charles P. Bianchi. Born and raised in Italy, he became known in the early 1920s as a swindler in North America for his money-making scheme. He promised clients a 50% profit within 45 days or 100% profit within 90 days, by buying discounted postal reply coupons in other countries and redeeming them at face value in the United States as a form of arbitrage. In reality, Ponzi was paying earlier investors using the investments of later investors. While this type of fraudulent investment scheme was not originally invented by Ponzi, it became so identified with him that it now is referred to as a "Ponzi scheme". His scheme ran for over a year before it collapsed, costing his "investors" $20 million.

Exam Probability: **Medium**

22. *Answer choices:*

(see index for correct answer)

- a. David G. Friehling
- b. Airplane game
- c. Charles Ponzi
- d. Valerie Red-Horse

Guidance: level 1

:: ::

_____ in the United States is a federal and state program that helps with medical costs for some people with limited income and resources. _____ also offers benefits not normally covered by Medicare, including nursing home care and personal care services. The Health Insurance Association of America describes _____ as "a government insurance program for persons of all ages whose income and resources are insufficient to pay for health care." _____ is the largest source of funding for medical and health-related services for people with low income in the United States, providing free health insurance to 74 million low-income and disabled people as of 2017. It is a means-tested program that is jointly funded by the state and federal governments and managed by the states, with each state currently having broad leeway to determine who is eligible for its implementation of the program. States are not required to participate in the program, although all have since 1982. _____ recipients must be U.S. citizens or qualified non-citizens, and may include low-income adults, their children, and people with certain disabilities. Poverty alone does not necessarily qualify someone for _____ .

Exam Probability: **Low**

23. *Answer choices:*

(see index for correct answer)

- a. Medicaid
- b. Character
- c. hierarchical perspective
- d. interpersonal communication

Guidance: level 1

_____ is a product prepared from the leaves of the _____ plant by curing them. The plant is part of the genus Nicotiana and of the Solanaceae family. While more than 70 species of _____ are known, the chief commercial crop is N. tabacum. The more potent variant N. rustica is also used around the world.

Exam Probability: **High**

24. *Answer choices:*

(see index for correct answer)

- a. similarity-attraction theory
- b. open system
- c. Sarbanes-Oxley act of 2002
- d. corporate values

Guidance: level 1

:: Types of marketing ::

_____ is an advertisement strategy in which a company uses surprise and/or unconventional interactions in order to promote a product or service. It is a type of publicity. The term was popularized by Jay Conrad Levinson's 1984 book _____ .

Exam Probability: **High**

25. *Answer choices:*

(see index for correct answer)

- a. Customerization
- b. Diversity marketing
- c. Figure of merit
- d. Vertical integration

Guidance: level 1

:: ::

In ecology, a _____ is the type of natural environment in which a particular species of organism lives. It is characterized by both physical and biological features. A species' _____ is those places where it can find food, shelter, protection and mates for reproduction.

Exam Probability: **Medium**

26. *Answer choices:*

(see index for correct answer)

- a. co-culture
- b. Character
- c. Habitat
- d. Sarbanes-Oxley act of 2002

Guidance: level 1

:: False advertising law ::

The Lanham Act is the primary federal trademark statute of law in the United States. The Act prohibits a number of activities, including trademark infringement, trademark dilution, and false advertising.

Exam Probability: **High**

27. *Answer choices:*

(see index for correct answer)

- a. Rebecca Tushnet
- b. Lanham Act

Guidance: level 1

:: Corporate scandals ::

The _____ was a privately held international group of financial services companies controlled by Allen Stanford, until it was seized by United States authorities in early 2009. Headquartered in the Galleria Tower II in Uptown Houston, Texas, it had 50 offices in several countries, mainly in the Americas, included the Stanford International Bank, and said it managed US$8.5 billion of assets for more than 30,000 clients in 136 countries on six continents. On February 17, 2009, U.S. Federal agents placed the company into receivership due to charges of fraud. Ten days later, the U.S. Securities and Exchange Commission amended its complaint to accuse Stanford of turning the company into a "massive Ponzi scheme".

Exam Probability: **Medium**

28. *Answer choices:*

(see index for correct answer)

- a. Crawford Texas Peace House
- b. Harken Energy scandal
- c. Stanford Financial Group
- d. Madoff investment scandal

Guidance: level 1

:: Writs ::

In common law, a writ of _____ is a writ whereby a private individual who assists a prosecution can receive all or part of any penalty imposed. Its name is an abbreviation of the Latin phrase _____ pro domino rege quam pro se ipso in hac parte sequitur, meaning "[he] who sues in this matter for the king as well as for himself."

Exam Probability: **Low**

29. *Answer choices:*

(see index for correct answer)

- a. Writ of assistance
- b. Writ of execution

Guidance: level 1

:: ::

The Catholic Church, also known as the Roman Catholic Church, is the largest Christian church, with approximately 1.3 billion baptised Catholics worldwide as of 2017. As the world's oldest continuously functioning international institution, it has played a prominent role in the history and development of Western civilisation. The church is headed by the Bishop of Rome, known as the pope. Its central administration, the Holy See, is in the Vatican City, an enclave within the city of Rome in Italy.

Exam Probability: **Low**

30. *Answer choices:*

(see index for correct answer)

- a. levels of analysis
- b. open system
- c. Sarbanes-Oxley act of 2002
- d. imperative

Guidance: level 1

A _____ service is an online platform which people use to build social networks or social relationship with other people who share similar personal or career interests, activities, backgrounds or real-life connections.

Exam Probability: **High**

31. *Answer choices:*

(see index for correct answer)

- a. Character
- b. co-culture
- c. process perspective
- d. empathy

Guidance: level 1

:: Labour relations ::

_____ is a field of study that can have different meanings depending on the context in which it is used. In an international context, it is a subfield of labor history that studies the human relations with regard to work – in its broadest sense – and how this connects to questions of social inequality. It explicitly encompasses unregulated, historical, and non-Western forms of labor. Here, _____ define "for or with whom one works and under what rules. These rules determine the type of work, type and amount of remuneration, working hours, degrees of physical and psychological strain, as well as the degree of freedom and autonomy associated with the work."

Exam Probability: **High**

32. *Answer choices:*
(see index for correct answer)

- a. Impasse
- b. Union representative
- c. Labor relations
- d. Boulwarism

Guidance: level 1

:: Human resource management ::

_____ is the ethics of an organization, and it is how an organization responds to an internal or external stimulus. _____ is interdependent with the organizational culture. Although it is akin to both organizational behavior and industrial and organizational psychology as well as business ethics on the micro and macro levels, _____ is neither OB or I/O psychology, nor is it solely business ethics . _____ express the values of an organization to its employees and/or other entities irrespective of governmental and/or regulatory laws.

Exam Probability: **Medium**

33. *Answer choices:*

(see index for correct answer)

- a. Adaptive performance
- b. ABC Consultants
- c. human resource
- d. Broadbanding

Guidance: level 1

:: Corporations law ::

A normal _____ consists of various departments that contribute to the company's overall mission and goals. Common departments include Marketing, [Finance, [[Operations managementOperations, Human Resource, and IT. These five divisions represent the major departments within a publicly traded company, though there are often smaller departments within autonomous firms. There is typically a CEO, and Board of Directors composed of the directors of each department. There are also company presidents, vice presidents, and CFOs.There is a great diversity in corporate forms as enterprises may range from single company to multi-corporate conglomerate. The four main _____ s are Functional, Divisional, Geographic, and the Matrix.Realistically, most corporations tend to have a "hybrid" structure, which is a combination of different models with one dominant strategy.

Exam Probability: **Medium**

34. *Answer choices:*

(see index for correct answer)

- a. Direct debit dividend contributions
- b. Corporate law
- c. Comply or explain
- d. Corporate structure

Guidance: level 1

:: Corporate governance ::

_____ refers to the practice of members of a corporate board of directors serving on the boards of multiple corporations. A person that sits on multiple boards is known as a multiple director. Two firms have a direct interlock if a director or executive of one firm is also a director of the other, and an indirect interlock if a director of each sits on the board of a third firm. This practice, although widespread and lawful, raises questions about the quality and independence of board decisions.

Exam Probability: **High**

35. *Answer choices:*

(see index for correct answer)

- a. Board-only
- b. Corporate title
- c. Interlocking directorate
- d. Corporate security

Guidance: level 1

:: ::

_____ is the introduction of contaminants into the natural environment that cause adverse change. _____ can take the form of chemical substances or energy, such as noise, heat or light. Pollutants, the components of _____, can be either foreign substances/energies or naturally occurring contaminants. _____ is often classed as point source or nonpoint source _____. In 2015, _____ killed 9 million people in the world.

Exam Probability: **Medium**

36. *Answer choices:*

(see index for correct answer)

- a. corporate values
- b. process perspective
- c. deep-level diversity
- d. Pollution

Guidance: level 1

:: Professional ethics ::

In the mental health field, a _____ is a situation where multiple roles exist between a therapist, or other mental health practitioner, and a client. _____ s are also referred to as multiple relationships, and these two terms are used interchangeably in the research literature. The American Psychological Association Ethical Principles of Psychologists and Code of Conduct is a resource that outlines ethical standards and principles to which practitioners are expected to adhere. Standard 3.05 of the APA ethics code outlines the definition of multiple relationships. Dual or multiple relationships occur when.

Exam Probability: **High**

37. *Answer choices:*

(see index for correct answer)

- a. ethical code
- b. Continuous professional development
- c. professional conduct

Guidance: level 1

:: ::

_____ generally refers to a focus on the needs or desires of one's self. A number of philosophical, psychological, and economic theories examine the role of _____ in motivating human action.

Exam Probability: **Low**

38. *Answer choices:*

(see index for correct answer)

- a. surface-level diversity
- b. Self-interest
- c. imperative
- d. hierarchical

Guidance: level 1

:: Financial regulatory authorities of the United States ::

The _____ is an agency of the United States government responsible for consumer protection in the financial sector. CFPB's jurisdiction includes banks, credit unions, securities firms, payday lenders, mortgage-servicing operations, foreclosure relief services, debt collectors and other financial companies operating in the United States.

Exam Probability: **High**

39. *Answer choices:*

(see index for correct answer)

- a. Municipal Securities Rulemaking Board
- b. Internal Revenue Service
- c. Commodity Futures Trading Commission
- d. Federal Deposit Insurance Corporation

Guidance: level 1

:: ::

_____ is the study and management of exchange relationships. _____ is the business process of creating relationships with and satisfying customers. With its focus on the customer, _____ is one of the premier components of business management.

Exam Probability: **High**

40. Answer choices:

(see index for correct answer)

- a. process perspective
- b. hierarchical
- c. similarity-attraction theory
- d. Marketing

Guidance: level 1

:: Monopoly (economics) ::

A _____ is a form of intellectual property that gives its owner the legal right to exclude others from making, using, selling, and importing an invention for a limited period of years, in exchange for publishing an enabling public disclosure of the invention. In most countries _____ rights fall under civil law and the _____ holder needs to sue someone infringing the _____ in order to enforce his or her rights. In some industries _____ s are an essential form of competitive advantage; in others they are irrelevant.

Exam Probability: **Medium**

41. Answer choices:

(see index for correct answer)

- a. Legal monopoly
- b. Monopoly

- c. Patent
- d. Eisenkammer Pirna

Guidance: level 1

:: ::

In regulatory jurisdictions that provide for it, _____ is a group of laws and organizations designed to ensure the rights of consumers as well as fair trade, competition and accurate information in the marketplace. The laws are designed to prevent the businesses that engage in fraud or specified unfair practices from gaining an advantage over competitors. They may also provides additional protection for those most vulnerable in society. _____ laws are a form of government regulation that aim to protect the rights of consumers. For example, a government may require businesses to disclose detailed information about products—particularly in areas where safety or public health is an issue, such as food.

Exam Probability: **High**

42. *Answer choices:*

(see index for correct answer)

- a. Consumer Protection
- b. Character
- c. functional perspective
- d. surface-level diversity

Guidance: level 1

:: Ethical banking ::

A _____ or community development finance institution - abbreviated in both cases to CDFI - is a financial institution that provides credit and financial services to underserved markets and populations, primarily in the USA but also in the UK. A CDFI may be a community development bank, a community development credit union , a community development loan fund , a community development venture capital fund , a microenterprise development loan fund, or a community development corporation.

Exam Probability: **High**

43. *Answer choices:*

(see index for correct answer)

- a. ShoreBank
- b. Community development financial institution
- c. Shared Interest
- d. Cultura Sparebank

Guidance: level 1

:: Separation of investment and commercial banking ::

The _____ refers to § 619 of the Dodd–Frank Wall Street Reform and Consumer Protection Act. The rule was originally proposed by American economist and former United States Federal Reserve Chairman Paul Volcker to restrict United States banks from making certain kinds of speculative investments that do not benefit their customers. Volcker argued that such speculative activity played a key role in the financial crisis of 2007–2008. The rule is often referred to as a ban on proprietary trading by commercial banks, whereby deposits are used to trade on the bank's own accounts, although a number of exceptions to this ban were included in the Dodd-Frank law.

Exam Probability: **Medium**

44. *Answer choices:*

(see index for correct answer)

- a. Bank Holding Company Act
- b. Commercial bank
- c. GLBA
- d. Volcker Rule

Guidance: level 1

:: Environmental economics ::

_____ is an institutional arrangement designed to help producers in developing countries achieve better trading conditions. Members of the _____ movement advocate the payment of higher prices to exporters, as well as improved social and environmental standards. The movement focuses in particular on commodities, or products which are typically exported from developing countries to developed countries, but also consumed in domestic markets most notably handicrafts, coffee, cocoa, wine, sugar, fresh fruit, chocolate, flowers and gold. The movement seeks to promote greater equity in international trading partnerships through dialogue, transparency, and respect. It promotes sustainable development by offering better trading conditions to, and securing the rights of, marginalized producers and workers in developing countries. _____ is grounded in three core beliefs; first, producers have the power to express unity with consumers. Secondly, the world trade practices that currently exist promote the unequal distribution of wealth between nations. Lastly, buying products from producers in developing countries at a fair price is a more efficient way of promoting sustainable development than traditional charity and aid.

Exam Probability: **High**

45. *Answer choices:*

(see index for correct answer)

- a. Fair trade
- b. Loan closet
- c. Eco-investing
- d. Ecolabel

Guidance: level 1

:: Utilitarianism ::

_____ is a family of consequentialist ethical theories that promotes actions that maximize happiness and well-being for the majority of a population. Although different varieties of _____ admit different characterizations, the basic idea behind all of them is to in some sense maximize utility, which is often defined in terms of well-being or related concepts. For instance, Jeremy Bentham, the founder of _____, described utility as

Exam Probability: **Low**

46. *Answer choices:*

(see index for correct answer)

- a. Hedonism
- b. Utilitarian bioethics
- c. The Methods of Ethics
- d. Consequentialism

Guidance: level 1

:: Timber industry ::

The _____ is an international non-profit, multi-stakeholder organization established in 1993 to promote responsible management of the world's forests. The FSC does this by setting standards on forest products, along with certifying and labeling them as eco-friendly.

Exam Probability: **High**

47. *Answer choices:*

(see index for correct answer)

- a. National Hardwood Lumber Association
- b. Brettstapel
- c. Baron of Renfrew
- d. Forest Stewardship Council

Guidance: level 1

:: Ethically disputed business practices ::

_____ is the trading of a public company's stock or other securities by individuals with access to nonpublic information about the company. In various countries, some kinds of trading based on insider information is illegal. This is because it is seen as unfair to other investors who do not have access to the information, as the investor with insider information could potentially make larger profits than a typical investor could make. The rules governing _____ are complex and vary significantly from country to country. The extent of enforcement also varies from one country to another. The definition of insider in one jurisdiction can be broad, and may cover not only insiders themselves but also any persons related to them, such as brokers, associates and even family members. A person who becomes aware of non-public information and trades on that basis may be guilty of a crime.

Exam Probability: **High**

48. *Answer choices:*

(see index for correct answer)

- a. Banishment room
- b. Insider trading
- c. at-will
- d. Boiler room

Guidance: level 1

:: ::

_____ is the collection of mechanisms, processes and relations by which corporations are controlled and operated. Governance structures and principles identify the distribution of rights and responsibilities among different participants in the corporation and include the rules and procedures for making decisions in corporate affairs. _____ is necessary because of the possibility of conflicts of interests between stakeholders, primarily between shareholders and upper management or among shareholders.

Exam Probability: **Medium**

49. *Answer choices:*

(see index for correct answer)

- a. similarity-attraction theory
- b. personal values
- c. Character
- d. hierarchical

Guidance: level 1

:: Product certification ::

_____ is food produced by methods that comply with the standards of organic farming. Standards vary worldwide, but organic farming features practices that cycle resources, promote ecological balance, and conserve biodiversity. Organizations regulating organic products may restrict the use of certain pesticides and fertilizers in the farming methods used to produce such products. _____ s typically are not processed using irradiation, industrial solvents, or synthetic food additives.

Exam Probability: **Low**

50. *Answer choices:*
(see index for correct answer)

- a. California Certified Organic Farmers
- b. Organic Crop Improvement Association
- c. Organic food
- d. Product certification

Guidance: level 1

:: Socialism ::

_____ is a label used to define the first currents of modern socialist thought as exemplified by the work of Henri de Saint-Simon, Charles Fourier, Étienne Cabet and Robert Owen.

Exam Probability: **Medium**

51. *Answer choices:*

(see index for correct answer)

- a. Facilitation board
- b. Sexualization
- c. Market abolitionism
- d. History of socialism

Guidance: level 1

:: ::

_____ is the means to see, hear, or become aware of something or someone through our fundamental senses. The term _____ derives from the Latin word perceptio, and is the organization, identification, and interpretation of sensory information in order to represent and understand the presented information, or the environment.

Exam Probability: **Low**

52. *Answer choices:*

(see index for correct answer)

- a. hierarchical
- b. Perception
- c. cultural
- d. open system

Guidance: level 1

:: ::

_____ or accountancy is the measurement, processing, and communication of financial information about economic entities such as businesses and corporations. The modern field was established by the Italian mathematician Luca Pacioli in 1494. _____ , which has been called the "language of business", measures the results of an organization's economic activities and conveys this information to a variety of users, including investors, creditors, management, and regulators. Practitioners of _____ are known as accountants. The terms "_____" and "financial reporting" are often used as synonyms.

Exam Probability: **Low**

53. *Answer choices:*

(see index for correct answer)

- a. Character
- b. imperative
- c. functional perspective

- d. Accounting

Guidance: level 1

:: Data management ::

_____ is a form of intellectual property that grants the creator of an original creative work an exclusive legal right to determine whether and under what conditions this original work may be copied and used by others, usually for a limited term of years. The exclusive rights are not absolute but limited by limitations and exceptions to _____ law, including fair use. A major limitation on _____ on ideas is that _____ protects only the original expression of ideas, and not the underlying ideas themselves.

Exam Probability: **High**

54. *Answer choices:*
(see index for correct answer)

- a. Data auditing
- b. Holos
- c. Reference table
- d. Copyright

Guidance: level 1

:: ::

_____ Corporation was an American energy, commodities, and services company based in Houston, Texas. It was founded in 1985 as a merger between Houston Natural Gas and InterNorth, both relatively small regional companies. Before its bankruptcy on December 3, 2001, _____ employed approximately 29,000 staff and was a major electricity, natural gas, communications and pulp and paper company, with claimed revenues of nearly $101 billion during 2000. Fortune named _____ "America's Most Innovative Company" for six consecutive years.

Exam Probability: **Medium**

55. *Answer choices:*

(see index for correct answer)

- a. information systems assessment
- b. Enron
- c. levels of analysis
- d. cultural

Guidance: level 1

:: Majority–minority relations ::

It was established as axiomatic in anthropological research by Franz Boas in the first few decades of the 20th century and later popularized by his students. Boas first articulated the idea in 1887: "civilization is not something absolute, but ... is relative, and ... our ideas and conceptions are true only so far as our civilization goes". However, Boas did not coin the term.

Exam Probability: **Medium**

56. *Answer choices:*

(see index for correct answer)

- a. positive discrimination
- b. cultural dissonance
- c. Affirmative action

Guidance: level 1

:: Fraud ::

In the United States, _____ is the claiming of Medicare health care reimbursement to which the claimant is not entitled. There are many different types of _____ , all of which have the same goal: to collect money from the Medicare program illegitimately.

Exam Probability: **High**

57. *Answer choices:*

(see index for correct answer)

- a. Clothing scam companies
- b. Medicare fraud
- c. Fraud in the factum
- d. Welfare queen

Guidance: level 1

:: ::

The _____ Group is a global financial investment management and insurance company headquartered in Des Moines, Iowa.

Exam Probability: **Low**

58. *Answer choices:*

(see index for correct answer)

- a. cultural
- b. hierarchical perspective
- c. imperative
- d. co-culture

Guidance: level 1

:: ::

The _____ is an 1848 political pamphlet by the German philosophers Karl Marx and Friedrich Engels. Commissioned by the Communist League and originally published in London just as the Revolutions of 1848 began to erupt, the Manifesto was later recognised as one of the world's most influential political documents. It presents an analytical approach to the class struggle and the conflicts of capitalism and the capitalist mode of production, rather than a prediction of communism's potential future forms.

Exam Probability: **Low**

59. *Answer choices:*

(see index for correct answer)

- a. hierarchical perspective
- b. cultural
- c. personal values
- d. Communist Manifesto

Guidance: level 1

Accounting

Accounting or accountancy is the measurement, processing, and communication of financial information about economic entities such as businesses and corporations. The modern field was established by the Italian mathematician Luca Pacioli in 1494. Accounting, which has been called the "language of business", measures the results of an organization's economic activities and conveys this information to a variety of users, including investors, creditors, management, and regulators.

:: Accounting in the United States ::

The _____ is the source of generally accepted accounting principles used by state and local governments in the United States. As with most of the entities involved in creating GAAP in the United States, it is a private, non-governmental organization.

Exam Probability: **High**

1. *Answer choices:*

(see index for correct answer)

- a. Legal liability of certified public accountants
- b. Variable interest entity
- c. Governmental Accounting Standards Board
- d. International Qualification Examination

Guidance: level 1

:: Types of business entity ::

A sole _____, also known as the sole trader, individual entrepreneurship or _____, is a type of enterprise that is owned and run by one person and in which there is no legal distinction between the owner and the business entity. A sole trader does not necessarily work 'alone'—it is possible for the sole trader to employ other people.

Exam Probability: **High**

2. *Answer choices:*

(see index for correct answer)

- a. Proprietorship
- b. Open joint-stock company
- c. Value-added reseller
- d. Massachusetts business trust

Guidance: level 1

:: Marketing ::

_____ or stock is the goods and materials that a business holds for the ultimate goal of resale.

Exam Probability: **Low**

3. *Answer choices:*

(see index for correct answer)

- a. Brand
- b. Lifestyle brand
- c. Fifth screen
- d. Inventory

Guidance: level 1

:: ::

_____ is the process of making predictions of the future based on past and present data and most commonly by analysis of trends. A commonplace example might be estimation of some variable of interest at some specified future date. Prediction is a similar, but more general term. Both might refer to formal statistical methods employing time series, cross-sectional or longitudinal data, or alternatively to less formal judgmental methods. Usage can differ between areas of application: for example, in hydrology the terms "forecast" and "_____" are sometimes reserved for estimates of values at certain specific future times, while the term "prediction" is used for more general estimates, such as the number of times floods will occur over a long period.

Exam Probability: **Low**

4. *Answer choices:*

(see index for correct answer)

- a. deep-level diversity
- b. functional perspective
- c. Forecasting
- d. similarity-attraction theory

Guidance: level 1

:: Foreign exchange market ::

A currency, in the most specific sense is money in any form when in use or circulation as a medium of exchange, especially circulating banknotes and coins. A more general definition is that a currency is a system of money in common use, especially for people in a nation. Under this definition, US dollars, pounds sterling, Australian dollars, European euros, Russian rubles and Indian Rupees are examples of currencies. These various currencies are recognized as stores of value and are traded between nations in foreign exchange markets, which determine the relative values of the different currencies. Currencies in this sense are defined by governments, and each type has limited boundaries of acceptance.

Exam Probability: **Medium**

5. *Answer choices:*

(see index for correct answer)

- a. Foreign exchange company
- b. Monetary unit
- c. Avignon Exchange
- d. Trade-weighted US dollar index

Guidance: level 1

:: Free accounting software ::

A _____ is the principal book or computer file for recording and totaling economic transactions measured in terms of a monetary unit of account by account type, with debits and credits in separate columns and a beginning monetary balance and ending monetary balance for each account.

Exam Probability: **Low**

6. *Answer choices:*

(see index for correct answer)

- a. Frontaccounting
- b. GnuCash
- c. TurboCASH
- d. JFin

Guidance: level 1

:: Management accounting ::

_____ is the process of recording, classifying, analyzing, summarizing, and allocating costs associated with a process,after that developing various courses of action to control the costs. Its goal is to advise the management on how to optimize business practices and processes based on cost efficiency and capability. _____ provides the detailed cost information that management needs to control current operations and plan for the future.

Exam Probability: **Low**

7. *Answer choices:*

(see index for correct answer)

- a. RCA open-source application
- b. Cost accounting

- c. Cost driver
- d. Hedge accounting

Guidance: level 1

:: Accounting ::

It is the period for which books are balanced and the financial statements are prepared. Generally, the _____ consists of 12 months. However the beginning of the _____ differs according to the jurisdiction. For example, one entity may follow the regular calendar year, i.e. January to December as the accounting year, while another entity may follow April to March as the _____ .

Exam Probability: **High**

8. *Answer choices:*

(see index for correct answer)

- a. Merdiban
- b. Accounting period
- c. INPACT International
- d. Cost allocation

Guidance: level 1

:: Expense ::

_____ relates to the cost of borrowing money. It is the price that a lender charges a borrower for the use of the lender's money. On the income statement, _____ can represent the cost of borrowing money from banks, bond investors, and other sources. _____ is different from operating expense and CAPEX, for it relates to the capital structure of a company, and it is usually tax-deductible.

Exam Probability: **Medium**

9. *Answer choices:*

(see index for correct answer)

- a. Corporate travel
- b. Interest expense
- c. Expense account
- d. expenditure

Guidance: level 1

:: Actuarial science ::

The _____ is the greater benefit of receiving money now rather than an identical sum later. It is founded on time preference.

Exam Probability: **Medium**

10. *Answer choices:*

(see index for correct answer)

- a. Solvency ratio
- b. Time value of money
- c. Reliability theory
- d. Actuarial science

Guidance: level 1

:: ::

In accounting, the _____ is a measure of the number of times inventory is sold or used in a time period such as a year. It is calculated to see if a business has an excessive inventory in comparison to its sales level. The equation for _____ equals the cost of goods sold divided by the average inventory. _____ is also known as inventory turns, merchandise turnover, stockturn, stock turns, turns, and stock turnover.

Exam Probability: **Medium**

11. *Answer choices:*

(see index for correct answer)

- a. Sarbanes-Oxley act of 2002
- b. deep-level diversity
- c. functional perspective
- d. personal values

Guidance: level 1

:: Accounting software ::

_____ describes a type of application software that records and processes accounting transactions within functional modules such as accounts payable, accounts receivable, journal, general ledger, payroll, and trial balance. It functions as an accounting information system. It may be developed in-house by the organization using it, may be purchased from a third party, or may be a combination of a third-party application software package with local modifications. _____ may be on-line based, accessed anywhere at any time with any device which is Internet enabled, or may be desktop based. It varies greatly in its complexity and cost.

Exam Probability: **Low**

12. *Answer choices:*

(see index for correct answer)

- a. Gem Accounts
- b. Amortization calculator
- c. PyBookie
- d. Accounting software

Guidance: level 1

:: Television terminology ::

A nonprofit organization, also known as a non-business entity, _____ organization, or nonprofit institution, is dedicated to furthering a particular social cause or advocating for a shared point of view. In economic terms, it is an organization that uses its surplus of the revenues to further achieve its ultimate objective, rather than distributing its income to the organization's shareholders, leaders, or members. Nonprofits are tax exempt or charitable, meaning they do not pay income tax on the money that they receive for their organization. They can operate in religious, scientific, research, or educational settings.

Exam Probability: **High**

13. *Answer choices:*

(see index for correct answer)

- a. nonprofit
- b. multiplexing
- c. Not-for-profit
- d. distance learning

Guidance: level 1

:: ::

An _____ , for United States federal income tax, is a closely held corporation that makes a valid election to be taxed under Subchapter S of Chapter 1 of the Internal Revenue Code. In general, _____ s do not pay any income taxes. Instead, the corporation's income or losses are divided among and passed through to its shareholders. The shareholders must then report the income or loss on their own individual income tax returns.

Exam Probability: **High**

14. *Answer choices:*

(see index for correct answer)

- a. Character
- b. co-culture
- c. deep-level diversity
- d. S corporation

Guidance: level 1

:: ::

A _____ is an entity that owes a debt to another entity. The entity may be an individual, a firm, a government, a company or other legal person. The counterparty is called a creditor. When the counterpart of this debt arrangement is a bank, the _____ is more often referred to as a borrower.

Exam Probability: **Medium**

15. Answer choices:

(see index for correct answer)

- a. levels of analysis
- b. Debtor
- c. hierarchical perspective
- d. empathy

Guidance: level 1

:: Investment ::

_____ , and investment appraisal, is the planning process used to determine whether an organization's long term investments such as new machinery, replacement of machinery, new plants, new products, and research development projects are worth the funding of cash through the firm's capitalization structure . It is the process of allocating resources for major capital, or investment, expenditures. One of the primary goals of _____ investments is to increase the value of the firm to the shareholders.

Exam Probability: **Medium**

16. Answer choices:

(see index for correct answer)

- a. Investment function
- b. Self-invested personal pension
- c. Capital budgeting

- d. Shipping Investments

Guidance: level 1

:: Fundamental analysis ::

_____ is the monetary value of earnings per outstanding share of common stock for a company.

Exam Probability: **High**

17. *Answer choices:*
(see index for correct answer)

- a. Goldman Sachs asset management factor model
- b. Earnings per share
- c. Enterprise value
- d. Market value added

Guidance: level 1

:: Cash flow ::

In corporate finance, _____ or _____ to firm is a way of looking at a business's cash flow to see what is available for distribution among all the securities holders of a corporate entity. This may be useful to parties such as equity holders, debt holders, preferred stock holders, and convertible security holders when they want to see how much cash can be extracted from a company without causing issues to its operations.

Exam Probability: **Medium**

18. *Answer choices:*

(see index for correct answer)

- a. Cash flow hedge
- b. Discounted payback period
- c. Cash flow forecasting
- d. Free cash flow

Guidance: level 1

:: Taxation ::

_____ refers to instances where a taxpayer can delay paying taxes to some future period. In theory, the net taxes paid should be the same. Taxes can sometimes be deferred indefinitely, or may be taxed at a lower rate in the future, particularly for deferral of income taxes.

Exam Probability: **Medium**

19. *Answer choices:*

(see index for correct answer)

- a. Airport improvement fee
- b. Tax deferral
- c. Taxpayer receipt
- d. Value capture

Guidance: level 1

:: Management accounting ::

_____ is the process of reviewing and analyzing a company's financial statements to make better economic decisions to earn income in future. These statements include the income statement, balance sheet, statement of cash flows, notes to accounts and a statement of changes in equity . _____ is a method or process involving specific techniques for evaluating risks, performance, financial health, and future prospects of an organization.

Exam Probability: **Low**

20. *Answer choices:*

(see index for correct answer)

- a. Direct material total variance
- b. Customer profitability
- c. Accounting management
- d. Corporate travel management

Guidance: level 1

:: Taxation in the United States ::

Basis, as used in United States tax law, is the original cost of property, adjusted for factors such as depreciation. When property is sold, the taxpayer pays/ taxes on a capital gain/ that equals the amount realized on the sale minus the sold property's basis.

Exam Probability: **High**

21. *Answer choices:*

(see index for correct answer)

- a. The Law that Never Was
- b. Private letter ruling
- c. Endowment tax
- d. Cost basis

Guidance: level 1

:: Shareholders ::

A _____ is a payment made by a corporation to its shareholders, usually as a distribution of profits. When a corporation earns a profit or surplus, the corporation is able to re-invest the profit in the business and pay a proportion of the profit as a _____ to shareholders. Distribution to shareholders may be in cash or, if the corporation has a _____ reinvestment plan, the amount can be paid by the issue of further shares or share repurchase. When _____s are paid, shareholders typically must pay income taxes, and the corporation does not receive a corporate income tax deduction for the _____ payments.

Exam Probability: **High**

22. *Answer choices:*

(see index for correct answer)

- a. Dividend
- b. Australian Shareholders Association
- c. Stock dilution
- d. Shareholder resolution

Guidance: level 1

:: Commerce ::

Continuation of an entity as a _____ is presumed as the basis for financial reporting unless and until the entity's liquidation becomes imminent. Preparation of financial statements under this presumption is commonly referred to as the _____ basis of accounting. If and when an entity's liquidation becomes imminent, financial statements are prepared under the liquidation basis of accounting.

Exam Probability: **Medium**

23. *Answer choices:*

(see index for correct answer)

- a. Kiosk
- b. E-receipt
- c. Going concern
- d. Contingent payment sales

Guidance: level 1

A _____ is a form of public administration which, in a majority of contexts, exists as the lowest tier of administration within a given state. The term is used to contrast with offices at state level, which are referred to as the central government, national government, or federal government and also to supranational government which deals with governing institutions between states. _____ s generally act within powers delegated to them by legislation or directives of the higher level of government. In federal states, _____ generally comprises the third tier of government, whereas in unitary states, _____ usually occupies the second or third tier of government, often with greater powers than higher-level administrative divisions.

Exam Probability: **High**

24. *Answer choices:*

(see index for correct answer)

- a. hierarchical perspective
- b. Local government
- c. personal values
- d. interpersonal communication

Guidance: level 1

:: Financial statements ::

A Statement of changes in equity and similarly the statement of changes in owner's equity for a sole trader, statement of changes in partners' equity for a partnership, statement of changes in Shareholders' equity for a Company or statement of changes in Taxpayers' equity for Government financial statements is one of the four basic financial statements.

Exam Probability: **High**

25. *Answer choices:*

(see index for correct answer)

- a. Balance sheet
- b. quarterly report
- c. Statement of retained earnings
- d. Financial report

Guidance: level 1

:: Taxation ::

In a tax system, the _____ is the ratio at which a business or person is taxed. There are several methods used to present a _____ : statutory, average, marginal, and effective. These rates can also be presented using different definitions applied to a tax base: inclusive and exclusive.

Exam Probability: **Low**

26. *Answer choices:*

(see index for correct answer)

- a. Tax rate
- b. Tax refund
- c. Security deposit
- d. Hotchpot

Guidance: level 1

:: Business ethics ::

In accounting and in most Schools of economic thought, _____ is a rational and unbiased estimate of the potential market price of a good, service, or asset. It takes into account such objectivity factors as.

Exam Probability: **Medium**

27. *Answer choices:*

(see index for correct answer)

- a. MBA Oath
- b. Tone at the top
- c. Fair value
- d. Whistleblower

Guidance: level 1

:: Business ::

The seller, or the provider of the goods or services, completes a sale in response to an acquisition, appropriation, requisition or a direct interaction with the buyer at the point of sale. There is a passing of title of the item, and the settlement of a price, in which agreement is reached on a price for which transfer of ownership of the item will occur. The seller, not the purchaser typically executes the sale and it may be completed prior to the obligation of payment. In the case of indirect interaction, a person who sells goods or service on behalf of the owner is known as a _____ man or _____ woman or _____ person, but this often refers to someone selling goods in a store/shop, in which case other terms are also common, including _____ clerk, shop assistant, and retail clerk.

Exam Probability: **High**

28. *Answer choices:*

(see index for correct answer)

- a. Retail design
- b. Co-creation
- c. Sales
- d. Legal governance, risk management, and compliance

Guidance: level 1

:: Insurance terms ::

A _____ in the broadest sense is a natural person or other legal entity who receives money or other benefits from a benefactor. For example, the _____ of a life insurance policy is the person who receives the payment of the amount of insurance after the death of the insured.

Exam Probability: **Medium**

29. *Answer choices:*

(see index for correct answer)

- a. Beneficiary
- b. New business strain
- c. Outstanding claims reserves
- d. Death spiral

Guidance: level 1

:: Management accounting ::

_____ s are costs that change as the quantity of the good or service that a business produces changes. _____ s are the sum of marginal costs over all units produced. They can also be considered normal costs. Fixed costs and _____ s make up the two components of total cost. Direct costs are costs that can easily be associated with a particular cost object. However, not all _____ s are direct costs. For example, variable manufacturing overhead costs are _____ s that are indirect costs, not direct costs. _____ s are sometimes called unit-level costs as they vary with the number of units produced.

Exam Probability: **High**

30. *Answer choices:*

(see index for correct answer)

- a. Variable cost
- b. Throughput accounting
- c. Invested capital
- d. Total benefits of ownership

Guidance: level 1

:: Finance ::

_____ , in finance and accounting, means stated value or face value. From this come the expressions at par , over par and under par .

Exam Probability: **High**

31. *Answer choices:*

(see index for correct answer)

- a. Equivalent annual cost
- b. Par value
- c. In-house lending
- d. OnDeck

Guidance: level 1

:: Taxation ::

_____ refers to the base upon which an income tax system imposes tax. Generally, it includes some or all items of income and is reduced by expenses and other deductions. The amounts included as income, expenses, and other deductions vary by country or system. Many systems provide that some types of income are not taxable and some expenditures not deductible in computing _____. Some systems base tax on _____ of the current period, and some on prior periods. _____ may refer to the income of any taxpayer, including individuals and corporations, as well as entities that themselves do not pay tax, such as partnerships, in which case it may be called "net profit".

Exam Probability: **Medium**

32. *Answer choices:*

(see index for correct answer)

- a. Taxable income
- b. Kharaj
- c. Tax uncertainty
- d. User charge

Guidance: level 1

:: Stock market ::

_____ is a form of corporate equity ownership, a type of security. The terms voting share and ordinary share are also used frequently in other parts of the world; "_____" being primarily used in the United States. They are known as Equity shares or Ordinary shares in the UK and other Commonwealth realms. This type of share gives the stockholder the right to share in the profits of the company, and to vote on matters of corporate policy and the composition of the members of the board of directors.

Exam Probability: **Medium**

33. *Answer choices:*

(see index for correct answer)

- a. Bellwether
- b. Clientele effect
- c. Indirect finance
- d. Red herring prospectus

Guidance: level 1

:: Economics terminology ::

A corporation's share capital or _____ is the portion of a corporation's equity that has been obtained by the issue of shares in the corporation to a shareholder, usually for cash. "Share capital" may also denote the number and types of shares that compose a corporation's share structure.

Exam Probability: **Low**

34. *Answer choices:*

(see index for correct answer)

- a. payee
- b. Capital stock
- c. capital accumulation
- d. total revenue

Guidance: level 1

:: Generally Accepted Accounting Principles ::

_____, or non-current liabilities, are liabilities that are due beyond a year or the normal operation period of the company. The normal operation period is the amount of time it takes for a company to turn inventory into cash. On a classified balance sheet, liabilities are separated between current and _____ to help users assess the company's financial standing in short-term and long-term periods. _____ give users more information about the long-term prosperity of the company, while current liabilities inform the user of debt that the company owes in the current period. On a balance sheet, accounts are listed in order of liquidity, so _____ come after current liabilities. In addition, the specific long-term liability accounts are listed on the balance sheet in order of liquidity. Therefore, an account due within eighteen months would be listed before an account due within twenty-four months. Examples of _____ are bonds payable, long-term loans, capital leases, pension liabilities, post-retirement healthcare liabilities, deferred compensation, deferred revenues, deferred income taxes, and derivative liabilities.

Exam Probability: **Medium**

35. Answer choices:

(see index for correct answer)

- a. Operating income before depreciation and amortization
- b. Cash method of accounting
- c. Long-term liabilities
- d. Operating statement

Guidance: level 1

:: Manufacturing ::

_____ costs are all manufacturing costs that are related to the cost object but cannot be traced to that cost object in an economically feasible way.

Exam Probability: **Medium**

36. Answer choices:

(see index for correct answer)

- a. Obeya
- b. Build to order
- c. Manufacturing overhead
- d. Build to stock

Guidance: level 1

:: Accounting systems ::

In bookkeeping, a _____ statement is a process that explains the difference on a specified date between the bank balance shown in an organization's bank statement, as supplied by the bank and the corresponding amount shown in the organization's own accounting records.

Exam Probability: **Medium**

37. *Answer choices:*
(see index for correct answer)

- a. Bank reconciliation
- b. Standard accounting practice
- c. Waste book
- d. Single-entry bookkeeping

Guidance: level 1

:: Expense ::

An _____ , operating expenditure, operational expense, operational expenditure or opex is an ongoing cost for running a product, business, or system. Its counterpart, a capital expenditure , is the cost of developing or providing non-consumable parts for the product or system. For example, the purchase of a photocopier involves capex, and the annual paper, toner, power and maintenance costs represents opex. For larger systems like businesses, opex may also include the cost of workers and facility expenses such as rent and utilities.

Exam Probability: **High**

38. *Answer choices:*

(see index for correct answer)

- a. Operating expense
- b. Stock option expensing
- c. Freight expense
- d. Interest expense

Guidance: level 1

:: Financial accounting ::

_____ in accounting is the process of treating investments in associate companies. Equity accounting is usually applied where an investor entity holds 20–50% of the voting stock of the associate company. The investor records such investments as an asset on its balance sheet. The investor's proportional share of the associate company's net income increases the investment, and proportional payments of dividends decrease it. In the investor's income statement, the proportional share of the investor's net income or net loss is reported as a single-line item.

Exam Probability: **High**

39. *Answer choices:*

(see index for correct answer)

- a. Money measurement
- b. Equity method
- c. Carry
- d. Floating capital

Guidance: level 1

:: Management accounting ::

_____ , or dollar contribution per unit, is the selling price per unit minus the variable cost per unit. "Contribution" represents the portion of sales revenue that is not consumed by variable costs and so contributes to the coverage of fixed costs. This concept is one of the key building blocks of break-even analysis.

Exam Probability: **Medium**

40. *Answer choices:*

(see index for correct answer)

- a. Standard cost
- b. Customer profitability
- c. Contribution margin
- d. Semi-variable cost

Guidance: level 1

:: Partnerships ::

Articles of partnership is a voluntary contract between/among two or more persons to place their capital, labor, and skills into business, with the understanding that there will be a sharing of the profits and losses between/among partners. Outside of North America, it is normally referred to simply as a _____ .

Exam Probability: **Low**

41. *Answer choices:*

(see index for correct answer)

- a. Uniform Limited Partnership Act
- b. Revised Uniform Partnership Act

- c. Partnership agreement
- d. Revised Uniform Limited Partnership Act

Guidance: level 1

:: Basic financial concepts ::

In finance, maturity or _____ refers to the final payment date of a loan or other financial instrument, at which point the principal is due to be paid.

Exam Probability: **Low**

42. *Answer choices:*
(see index for correct answer)

- a. Leverage cycle
- b. Maturity date
- c. Eurodollar
- d. Future-oriented

Guidance: level 1

:: ::

A _____ is the period used by governments for accounting and budget purposes, which varies between countries. It is also used for financial reporting by business and other organizations. Laws in many jurisdictions require company financial reports to be prepared and published on an annual basis, but generally do not require the reporting period to align with the calendar year. Taxation laws generally require accounting records to be maintained and taxes calculated on an annual basis, which usually corresponds to the _____ used for government purposes. The calculation of tax on an annual basis is especially relevant for direct taxation, such as income tax. Many annual government fees—such as Council rates, licence fees, etc.—are also levied on a _____ basis, while others are charged on an anniversary basis.

Exam Probability: **High**

43. *Answer choices:*
(see index for correct answer)

- a. Fiscal year
- b. open system
- c. personal values
- d. imperative

Guidance: level 1

:: Business law ::

An _____ is a natural person, business, or corporation that provides goods or services to another entity under terms specified in a contract or within a verbal agreement. Unlike an employee, an _____ does not work regularly for an employer but works as and when required, during which time they may be subject to law of agency. _____ s are usually paid on a freelance basis. Contractors often work through a limited company or franchise, which they themselves own, or may work through an umbrella company.

Exam Probability: **High**

44. *Answer choices:*

(see index for correct answer)

- a. Unfair Commercial Practices Directive
- b. Enhanced use lease
- c. Board of directors
- d. TRIPS Agreement

Guidance: level 1

:: Options (finance) ::

A _____ bond is a type of bond that allows the issuer of the bond to retain the privilege of redeeming the bond at some point before the bond reaches its date of maturity. In other words, on the call date, the issuer has the right, but not the obligation, to buy back the bonds from the bond holders at a defined call price. Technically speaking, the bonds are not really bought and held by the issuer but are instead cancelled immediately.

Exam Probability: **Low**

45. *Answer choices:*

(see index for correct answer)

- a. LEAPS
- b. Contingent value rights
- c. Callable
- d. Cash or share option

Guidance: level 1

:: Accounting terminology ::

_____ or capital expense is the money a company spends to buy, maintain, or improve its fixed assets, such as buildings, vehicles, equipment, or land. It is considered a _____ when the asset is newly purchased or when money is used towards extending the useful life of an existing asset, such as repairing the roof.

Exam Probability: **Low**

46. *Answer choices:*

(see index for correct answer)

- a. Checkoff
- b. Double-entry accounting

- c. Accrual
- d. Capital expenditure

Guidance: level 1

:: Accounting source documents ::

A _____ or account statement is a summary of financial transactions which have occurred over a given period on a bank account held by a person or business with a financial institution.

Exam Probability: **Medium**

47. *Answer choices:*

(see index for correct answer)

- a. Bank statement
- b. Invoice
- c. Credit memorandum
- d. Air waybill

Guidance: level 1

:: ::

From an accounting perspective, _____ is crucial because _____ and _____ taxes considerably affect the net income of most companies and because they are subject to laws and regulations.

Exam Probability: **Medium**

48. *Answer choices:*

(see index for correct answer)

- a. cultural
- b. information systems assessment
- c. Payroll
- d. similarity-attraction theory

Guidance: level 1

:: Financial accounting ::

In macroeconomics and international finance, the _____ is one of two primary components of the balance of payments, the other being the current account. Whereas the current account reflects a nation's net income, the _____ reflects net change in ownership of national assets.

Exam Probability: **Low**

49. *Answer choices:*

(see index for correct answer)

- a. Capital account
- b. Hidden asset
- c. Accounting identity
- d. Commuted cash value

Guidance: level 1

:: Types of accounting ::

Various _____ systems are used by various public sector entities. In the United States, for instance, there are two levels of government which follow different accounting standards set forth by independent, private sector boards. At the federal level, the Federal Accounting Standards Advisory Board sets forth the accounting standards to follow. Similarly, there is the _____ Standards Board for state and local level government.

Exam Probability: **High**

50. *Answer choices:*

(see index for correct answer)

- a. Personal environmental impact accounting
- b. Sustainability accounting
- c. Product control

Guidance: level 1

:: Inventory ::

_____ is a system of inventory in which updates are made on a periodic basis. This differs from perpetual inventory systems, where updates are made as seen fit.

Exam Probability: **High**

51. *Answer choices:*

(see index for correct answer)

- a. Item-level tagging
- b. Inventory control problem
- c. Periodic inventory
- d. Inventory optimization

Guidance: level 1

:: Financial ratios ::

The _____ or dividend-price ratio of a share is the dividend per share, divided by the price per share. It is also a company's total annual dividend payments divided by its market capitalization, assuming the number of shares is constant. It is often expressed as a percentage.

Exam Probability: **Medium**

52. Answer choices:

(see index for correct answer)

- a. Days sales outstanding
- b. Greeks
- c. Dividend yield
- d. Diluted earnings per share

Guidance: level 1

:: Accounting in the United States ::

The _____ is a private-sector, nonprofit corporation created by the Sarbanes–Oxley Act of 2002 to oversee the audits of public companies and other issuers in order to protect the interests of investors and further the public interest in the preparation of informative, accurate and independent audit reports. The PCAOB also oversees the audits of broker-dealers, including compliance reports filed pursuant to federal securities laws, to promote investor protection. All PCAOB rules and standards must be approved by the U.S. Securities and Exchange Commission.

Exam Probability: **Medium**

53. Answer choices:

(see index for correct answer)

- a. Beta Alpha Psi
- b. Statements on Auditing Procedure

- c. Plug
- d. Variable interest entity

Guidance: level 1

:: Expense ::

An _____ is the right to reimbursement of money spent by employees for work-related purposes. Some common _____ s are: administrative expense, amortization expense, bad debt expense, cost of goods sold, depreciation expense, freight-out, income tax expense, insurance expense, interest expense, loss on disposal of plant assets, maintenance and repairs expense, rent expense, salaries and wages expense, selling expense, supplies expense and utilities expense.

Exam Probability: **High**

54. *Answer choices:*

(see index for correct answer)

- a. Interest expense
- b. Freight expense
- c. Tax expense
- d. Stock option expensing

Guidance: level 1

:: Credit cards ::

The _____ Company, also known as Amex, is an American multinational financial services corporation headquartered in Three World Financial Center in New York City. The company was founded in 1850 and is one of the 30 components of the Dow Jones Industrial Average. The company is best known for its charge card, credit card, and traveler's cheque businesses.

Exam Probability: **Medium**

55. *Answer choices:*

(see index for correct answer)

- a. North American Bancard
- b. Rail travel card
- c. American Express
- d. Barclaycard

Guidance: level 1

:: Auditing ::

_____, as defined by accounting and auditing, is a process for assuring of an organization's objectives in operational effectiveness and efficiency, reliable financial reporting, and compliance with laws, regulations and policies. A broad concept, _____ involves everything that controls risks to an organization.

Exam Probability: **Low**

56. *Answer choices:*

(see index for correct answer)

- a. Vouching
- b. Communication audit
- c. Audit planning
- d. Internal control

Guidance: level 1

:: ::

_____ is capital that is contributed to a corporation by investors by purchase of stock from the corporation, the primary market, not by purchase of stock in the open market from other stockholders. It includes share capital as well as additional _____ .

Exam Probability: **Medium**

57. *Answer choices:*

(see index for correct answer)

- a. open system
- b. interpersonal communication
- c. imperative

- d. Paid-in capital

Guidance: level 1

:: Stock market ::

_____ is a form of stock which may have any combination of features not possessed by common stock including properties of both an equity and a debt instrument, and is generally considered a hybrid instrument. _____ s are senior to common stock, but subordinate to bonds in terms of claim and may have priority over common stock in the payment of dividends and upon liquidation. Terms of the _____ are described in the issuing company's articles of association or articles of incorporation.

Exam Probability: **Medium**

58. *Answer choices:*
(see index for correct answer)

- a. Thinkorswim
- b. London Stock Exchange Group
- c. Secondary market offering
- d. Split share corporation

Guidance: level 1

:: Auditing ::

An _____ is a security-relevant chronological record, set of records, and/or destination and source of records that provide documentary evidence of the sequence of activities that have affected at any time a specific operation, procedure, or event. Audit records typically result from activities such as financial transactions, scientific research and health care data transactions, or communications by individual people, systems, accounts, or other entities.

Exam Probability: **High**

59. *Answer choices:*

(see index for correct answer)

- a. Continuous auditing
- b. Auditing Standards Board
- c. Recovery Auditing
- d. Legal auditing

Guidance: level 1

INDEX: Correct Answers

Foundations of Business

1. a: Audience

2. d: Question

3. : Employment

4. : Interview

5. c: Contract

6. b: Shareholders

7. d: Accounts receivable

8. b: Restructuring

9. a: Health

10. d: Gross domestic product

11. b: Business process

12. d: Strategic alliance

13. : Cultural

14. c: Corporation

15. : Trade

16. b: Board of directors

17. d: Strategic planning

18. d: Explanation

19. : Performance

20. : Empowerment

21. a: System

22. b: Perception

23. d: Exchange rate

24. : Focus group

25. c: Political risk

26. b: Security

27. : Best practice

28. b: Benchmarking

29. a: Entrepreneurship

30. : Goal

31. c: Foreign direct investment

32. d: Money

33. a: Stock exchange

34. : Venture capital

35. d: Procurement

36. d: Free trade

37. a: Quality control

38. d: Balanced scorecard

39. c: Decision-making

40. c: Economic Development

41. c: Solution

42. d: Project management

43. b: Target market

44. c: Cash flow

45. c: Import

46. d: Office

47. : Initiative

48. c: Exercise

49. : Scheduling

50. b: Pattern

51. b: Property

52. d: Organizational structure

53. d: Inflation

54. c: Interest rate

55. d: Fraud

56. a: Accounting

57. a: Dimension

58. : Capital market

59. d: Buyer

Management

1. b: Hotel

2. : Meeting

3. : Supervisor

4. b: Free trade

5. : Review

6. d: Small business

7. b: Integrity

8. : Glass ceiling

9. b: Cooperative

10. : Empowerment

11. d: Entrepreneurship

12. a: Utility

13. a: Overtime

14. : Criticism

15. c: Variable cost

16. : Export

17. d: Argument

18. : Justice

19. d: Labor relations

20. : Budget

21. : Entrepreneur

22. : Sales

23. d: Lead

24. b: Task force

25. : Contingency theory

26. c: Reputation

27. a: Myers-Briggs type

28. b: Situational leadership

29. : Cost

30. d: Income

31. d: Product life cycle

32. : Absenteeism

33. d: Control chart

34. : Synergy

35. c: Protection

36. b: Management process

37. c: Linear programming

38. c: Inventory control

39. c: Problem

40. d: Halo effect

41. c: Process control

42. a: Frequency

43. b: Initiative

44. : Interview

45. b: Industry

46. c: Referent power

47. c: Restructuring

48. : Expatriate

49. b: Job design

50. : Coaching

51. c: Organizational learning

52. : Innovation

53. a: Questionnaire

54. b: Cost leadership

55. c: Quality management

56. a: Affirmative action

57. d: Property

58. a: Market research

59. d: Pension

Business law

1. d: Stock

2. d: Writ

3. c: Mortgage

4. c: Consumer protection

5. : Lease

6. c: Income

7. c: Warehouse receipt

8. d: Adverse possession

9. a: Security interest

10. d: Uniform Commercial Code

11. d: Misappropriation

12. a: Punitive

13. b: Foreign Corrupt Practices Act

14. d: Limited liability

15. d: Private law

16. d: Sole proprietorship

17. d: Merger

18. d: Holder in due course

19. b: Subsidiary

20. c: Contributory negligence

21. b: Supreme Court

22. a: Secured transaction

23. a: Accounting

24. b: Free trade

25. c: Directed verdict

26. a: Beneficiary

27. a: Incentive

28. b: False imprisonment

29. b: Hearing

30. a: Void contract

31. d: Duress

32. : Offeror

33. b: Federal question

34. c: De jure

35. d: Security agreement

36. c: Berne Convention

37. : Real property

38. b: Garnishment

39. b: Misrepresentation

40. : Criminal law

41. a: Unconscionability

42. c: Property

43. c: Estoppel

44. c: Contract

45. d: Inventory

46. d: Internal Revenue Service

47. : Jurisdiction

48. a: Security

49. : Specific performance

50. a: Relevant market

51. : Money laundering

52. d: Perfection

53. a: Deed

54. a: Foreclosure

55. b: Fair use

56. c: Lanham Act

57. : Breach of contract

58. : National Labor Relations Board

59. d: Respondeat superior

Finance

1. a: Manufacturing cost

2. c: Financial ratio

3. b: Asset

4. a: Futures contract

5. d: Annual report

6. b: Accounts receivable

7. d: Mutual fund

8. : Land

9. b: Net income

10. c: INDEX

11. b: Yield curve

12. : Restructuring

13. d: Initial public offering

14. d: Deferral

15. b: Manufacturing

16. d: Call option

17. d: Standard cost

18. d: Financial Accounting Standards Board

19. c: Raw material

20. c: Currency

21. d: Normal balance

22. a: Partnership

23. b: Dividend yield

24. c: Comprehensive income

25. : Purchasing

26. : Fiscal year

27. d: Preference

28. : Internal rate of return

29. c: Accountant

30. c: Schedule

31. a: Presentation

32. a: Financial market

33. d: Budget

34. : Fraud

35. b: Standard deviation

36. b: Forecasting

37. b: Commercial paper

38. b: Property

39. : Loan

40. d: Accounting period

41. c: Underwriting

42. c: Indenture

43. d: Preferred stock

44. : Break-even

45. b: Capital gain

46. c: Price

47. c: Operating expense

48. d: Retained earnings

49. : Face

50. c: Balanced scorecard

51. b: Internal Revenue Service

52. b: Arbitrage

53. : Financial risk

54. : Vacation

55. c: Cash flow

56. d: Intangible asset

57. a: Sinking fund

58. c: Sales

59. a: Time value of money

Human resource management

1. b: Affirmative action

2. d: Interview

3. d: Reinforcement

4. b: Aptitude

5. c: Phantom stock

6. d: Mission statement

7. a: Living wage

8. c: National Association of Colleges and Employers

9. a: Offshoring

10. b: Expatriate

11. a: Job sharing

12. b: Health Reimbursement Account

13. c: Task force

14. d: Total Reward

15. a: Glass ceiling

16. d: Unemployment benefits

17. : Disability insurance

18. a: Labor force

19. d: Questionnaire

20. c: Balance sheet

21. d: Halo effect

22. a: Payroll

23. : Executive officer

24. b: Needs analysis

25. a: Independent contractor

26. d: Self-actualization

27. a: Employment

28. c: Performance management

29. b: Employee stock ownership plan

30. b: Minimum wage

31. b: Structured interview

32. : Physician

33. c: Cost

34. c: Social loafing

35. b: Human resources

36. d: Job evaluation

37. d: Eustress

38. : Local union

39. b: Compensation and benefits

40. : New Deal

41. : Card check

42. c: Free agent

43. a: On-the-job training

44. b: Job satisfaction

45. c: Workplace violence

46. : Recruitment

47. c: Business game

48. : Profession

49. d: Predictive validity

50. b: Balanced scorecard

51. a: Cost of living

52. : Vesting

53. a: Professional employer organization

54. b: Job rotation

55. d: Delayering

56. : Transformational leadership

57. b: Internship

58. a: Knowledge worker

59. a: Impression management

Information systems

1. d: American Express

2. d: Mobile computing

3. d: Spyware

4. c: Zynga

5. : Picasa

6. b: Outsourcing

7. d: Output device

8. : User interface

9. b: Census

10. d: Intranet

11. : Business process

12. b: Interview

13. a: Viral marketing

14. d: Netscape

15. a: Big data

16. a: Accessibility

17. d: Critical success factor

18. a: Wiki

19. b: Enterprise application

20. a: Master data management

21. c: Random access

22. d: Authorization

23. b: Privacy

24. c: Consumer-to-consumer

25. : BitTorrent

26. a: Diagram

27. b: Credit card

28. c: Payment system

29. : Mass customization

30. c: Content management

31. : Transaction processing

32. c: Text mining

33. : Authentication protocol

34. a: Geocoding

35. c: Dashboard

36. b: Web content

37. a: Business process management

38. c: Social network

39. : Artificial intelligence

40. d: Master data

41. d: Expert system

42. d: Chart

43. : Cookie

44. c: Facebook

45. d: Entity-relationship

46. b: Local Area Network

47. c: Change control

48. : Query language

49. d: PayPal

50. c: Google Calendar

51. d: Information flow

52. : Change management

53. c: Sustainable

54. a: Disaster recovery plan

55. c: Long tail

56. c: Data link

57. b: Metadata

58. c: Disaster recovery

59. b: Enterprise search

Marketing

1. d: Comparative advertising

2. : Partnership

3. d: Brand management

4. c: Nonprofit

5. a: Merchandising

6. b: Product development

7. d: Contract

8. c: Commodity

9. d: Attention

10. c: Consumerism

11. d: Marketing strategy

12. d: Commerce

13. b: E-commerce

14. d: Standing

15. a: Market share

16. : Stock

17. d: Cost-plus pricing

18. b: Question

19. d: Customer satisfaction

20. b: Customer

21. b: Economies of scale

22. c: Publicity

23. a: Economy

24. d: Consultant

25. : Manager

26. d: Brand extension

27. : Secondary data

28. : Brand

29. d: Census

30. : Marketing communications

31. : Customer retention

32. c: Patent

33. b: Unique selling proposition

34. c: Asset

35. b: Qualitative research

36. d: Data collection

37. c: Competitor

38. a: Wall Street Journal

39. : Demand

40. : Evaluation

41. c: Intranet

42. : Blog

43. a: Price war

44. c: Technology

45. b: Intangibility

46. c: Small business

47. : Copyright

48. a: Business model

49. c: Strategic alliance

50. a: Telemarketing

51. d: Policy

52. a: Leadership

53. b: Market research

54. d: Interactive marketing

55. : Corporation

56. c: Marketing research

57. : Evolution

58. d: Quantitative research

59. b: Total Quality Management

Manufacturing

1. : Elastomer

2. d: Catalyst

3. c: Bullwhip effect

4. : Expediting

5. a: Water

6. d: Quality costs

7. b: Thomas Register

8. c: Asset

9. d: Cash register

10. b: Quality function deployment

11. b: Mary Kay

12. : Scheduling

13. : New product development

14. : Inventory

15. : Schedule

16. d: Workflow

17. : Cost reduction

18. : Distillation

19. d: Synergy

20. d: Reorder point

21. d: Voice of the customer

22. : Process flow diagram

23. d: Steering committee

24. c: Inventory control

25. b: Vendor relationship management

26. : Volume

27. c: Stakeholder management

28. b: Control limits

29. c: American Society for Quality

30. d: Remanufacturing

31. : Risk management

32. : Perfect competition

33. : Process capital

34. c: Lean manufacturing

35. b: Acceptance sampling

36. c: Process engineering

37. a: Lead

38. a: Zero Defects

39. d: Transaction cost

40. d: Waste

41. c: Information management

42. a: Purchase order

43. c: Kaizen

44. a: Service level

45. c: Process management

46. a: Downtime

47. : Quality control

48. b: Economic order quantity

49. : Pareto analysis

50. : Average cost

51. d: Cost driver

52. d: Indirect costs

53. c: Root cause

54. c: Aggregate planning

55. c: Original equipment manufacturer

56. b: Resource

57. c: Consensus

58. : Histogram

59. b: Technical support

Commerce

1. d: Bill of lading

2. a: Outsourcing

3. a: Household

4. b: Anticipation

5. a: Strategic alliance

6. c: Deregulation

7. a: Hearing

8. d: Jury

9. a: Investment

10. a: Fixed cost

11. : Liquidation

12. a: Asset

13. a: Standing

14. a: Trade

15. c: Freight forwarder

16. : Collaborative filtering

17. c: Boot

18. : Inflation

19. b: Supranational

20. : Encryption

21. : Budget

22. a: Strategic plan

23. a: E-procurement

24. a: Netflix

25. : Minimum wage

26. c: Personnel

27. : Purchasing

28. : Computer security

29. d: Wholesale

30. : Excite

31. a: Competitive advantage

32. c: Free market

33. d: Human resources

34. : Marketing

35. d: Organizational structure

36. : Stock

37. a: Broker

38. b: Marketspace

39. c: Information technology

40. d: Commodity

41. d: Short run

42. b: Automated Clearing House

43. b: Reverse auction

44. d: Insurance

45. : Import

46. b: Issuing bank

47. c: Vendor

48. b: Overtime

49. b: Tangible

50. d: Auction

51. : Phishing

52. c: Customer satisfaction

53. c: Interest

54. d: Empowerment

55. c: Good

56. c: Market segmentation

57. c: Marketing strategy

58. a: Lease

59. : Consignee

Business ethics

1. b: Great Depression

2. b: Model Rules of Professional Conduct

3. c: Transformational leadership

4. b: Aristotle

5. c: Six Sigma

6. b: Antitrust

7. d: Clean Water Act

8. d: New Deal

9. : Pyramid scheme

10. c: Workplace bullying

11. a: Hedonism

12. c: Empowerment

13. b: Corporate citizenship

14. b: Human nature

15. c: Reputation

16. a: Financial Stability Oversight Council

17. : Real estate

18. d: Sustainability

19. a: Solar power

20. c: Clayton Act

21. : Oil spill

22. c: Charles Ponzi

23. a: Medicaid

24. : Tobacco

25. : Guerrilla Marketing

26. c: Habitat

27. b: Lanham Act

28. c: Stanford Financial Group

29. c: Qui tam

30. : Catholicism

31. : Social networking

32. c: Labor relations

33. : Organizational ethics

34. d: Corporate structure

35. c: Interlocking directorate

36. d: Pollution

37. d: Dual relationship

38. b: Self-interest

39. : Consumer Financial Protection Bureau

40. d: Marketing

41. c: Patent

42. a: Consumer Protection

43. b: Community development financial institution

44. d: Volcker Rule

45. a: Fair trade

46. : Utilitarianism

47. d: Forest Stewardship Council

48. b: Insider trading

49. : Corporate governance

50. c: Organic food

51. : Utopian socialism

52. b: Perception

53. d: Accounting

54. d: Copyright

55. b: Enron

56. d: Cultural relativism

57. b: Medicare fraud

58. : Principal Financial

59. d: Communist Manifesto

Accounting

1. c: Governmental Accounting Standards Board

2. a: Proprietorship

3. d: Inventory

4. c: Forecasting

5. b: Monetary unit

6. : Ledger

7. b: Cost accounting

8. b: Accounting period

9. b: Interest expense

10. b: Time value of money

11. : Inventory turnover

12. d: Accounting software

13. c: Not-for-profit

14. d: S corporation

15. b: Debtor

16. c: Capital budgeting

17. b: Earnings per share

18. d: Free cash flow

19. b: Tax deferral

20. : Financial statement analysis

21. d: Cost basis

22. a: Dividend

23. c: Going concern

24. b: Local government

25. c: Statement of retained earnings

26. a: Tax rate

27. c: Fair value

28. c: Sales

29. a: Beneficiary

30. a: Variable cost

31. b: Par value

32. a: Taxable income

33. : Common stock

34. b: Capital stock

35. c: Long-term liabilities

36. c: Manufacturing overhead

37. a: Bank reconciliation

38. a: Operating expense

39. b: Equity method

40. c: Contribution margin

41. c: Partnership agreement

42. b: Maturity date

43. a: Fiscal year

44. : Independent contractor

45. c: Callable

46. d: Capital expenditure

47. a: Bank statement

48. c: Payroll

49. a: Capital account

50. d: Governmental accounting

51. c: Periodic inventory

52. c: Dividend yield

53. : Public Company Accounting Oversight Board

54. : Expense account

55. c: American Express

56. d: Internal control

57. d: Paid-in capital

58. : Preferred stock

59. : Audit trail

CPSIA information can be obtained
at www.ICGtesting.com
Printed in the USA
LVHW012248291019
635715LV00007B/562